Breeding and Hybridization of Food Fishes

The Author

Dr. K.P. Biswas, M.Sc, Ph.D, D.F.Sc. (Bombay), E.F. (West Germany), F.Z.S, F.A.B.S. (Kolkata), took interest in induced fish breeding in 1957 looking at the success of Dr. Hiralal Chaudhuri in Angul Fish farm in Odisha. In 1960, the author succeeded to induce a pair of mrigal (Indian major carp) in Balasore, Odisha by injecting fish pituitary hormone. Thereafter from 1961 onwards he induced a number of Indian major carps for breeding in the field and in controlled temperature in the laboratory. He also succeeded to ovulate female rohu by stimulating pituitary gland electrically with micro electrodes. In 1982, the author, as Director Fisheries, Andaman and Nicobar Admn., for the first time bred major carps to produce carp seed in Port Blair, Andaman Island. Even after his retirement from Govt. service, he bred major carps in Sagar Island, West Bengal and produced major carp seed locally in 2002. All these practical experiences gave him the impetus to write the book, Breeding and Hybridization of Food Fishes".

Breeding and Hybridization of Food Fishes

Dr. K.P. Biswas

2017

Daya Publishing House®

A Division of

Astral International Pvt. Ltd.

New Delhi – 110 002

Cataloging in Publication Data--DK
Courtesy: D.K. Agencies (P) Ltd. <docinfo@dkagencies.com>

Biswas, K. P. (Kamakhya Pada), 1936- **author.**
Breeding and hybridization of food fishes / author, Dr. K.P. Biswas.
   pages cm
Includes bibliographical references and index.
ISBN 978-93-86071-35-4 (International Edition)

1. Fishes--Breeding--India. 2. Fishes--Breeding. 3. Fishes--Hybridization--India. 4. Fishes--Hybridization. 5. Fish as food--India. 6. Fish as food. I. Title.
SH155.5.B57 2017   DDC 639.3   23

Published by   : Daya Publishing House®
                             A Division of
                             Astral International Pvt. Ltd.
                             – ISO 9001:2015 Certified Company –
                             4736/23, Ansari Road, Darya Ganj
                             New Delhi-110 002
                             Ph. 011-43549197, 23278134
                             E-mail: info@astralint.com
                             Website: www.astralint.com

— Dedication —

**Dedicated to Mrs. Manju Biswas
for Encouragement in
Writing the Book**

Acknowledgement

The author deeply acknowledge the help of Dr. N.A. Talwar for preparing the print out of hard copy of the manuscript.

K.P. Biswas

Preface

A news by a staff reporter of Statesman was published on 8th July, 1960, where artificial breeding of fish was reported to have been planned by the West Bengal Government at Kanchrapara Research Center. The news further elucidated that fish breeding by pituitary injection was first done in India by Dr. H. L. Chaudhuri of Union Government's Fisheries Department two years ago. The experiments were carried out at Cuttack on major carps. Dr. L. S. Ramsawami of Central College at Bangalore also made similar experiments on catfish.

The news further added that Dr. K. P. Biswas (the author), the then District Fishery Officer, Balasore, Odisha experimented with two pairs of mrigal (*Cirrhinus mrigala*) fish. Dr. Biswas injected pituitary hormone into two pairs of fish on July 2. The next day, the two pairs bred, fertilization being found to be 95 percent and hatching 80 percent. The total number of spawn was found to be over one lakh. Fish generally breed within 18 hours of the injection.

Over the news, Amritabazar Patrika dated 11th July, 1960 published a column criticising the figure of spawn production as "Fishful thinking" and expressed doubts about the correct counting of such fabulous spawn production reported by the District Fishery Officer. In the same newspaper dated 17th July, 1960, " asuda " in his column "Do you think" expressed his doubt to the readers about such huge number of offspring from a pair of mrigal.

This was the beginning of the involvement of the author with breeding of cultivated fishes, 54 years back; especially the major carps, which do not breed in confined ater and breed only in flowing rivers at monsoon, during June to August. The author did induced breeding of major carps, both experimentally in the laboratory under controlled temperature and in the field commercially from 1960 to 1963 at Cuttack, Odisha and there after in 1982 to 1983 at Port Blair, Andaman Islands for the first time in the Island. At the start, both in Balasore district and Port

Blair, the author had to do it by himself by making some make shift arrangements, as minimum infrastructure facilities were not available in both the places at that time.

Again after retirement from Government service in 1994, the author in 2002 conducted the induced breeding of major carps and constructed and operated one unit of circular hatchery in the premises of Prof. Amalesh Chaudhury's Marine Biological Research Station at Sagar Island at his request.

Besides, the author had conducted a number of experiments on induced breeding of rahu (*Labeo rohita*) in the year 1962 at Cuttack by stimulating the pituitary gland of the fish by electric shock with positive results of ovulation. The technique could not be standardized due to the author's departure to West Germany for electrophysiological work. Some of these experiments led the author to believe that suitable environmental stimulation (water temperature, rains etc.) can also trigger the breeding of major carps, as in case of bundh breeding, without application of gonadotropic hormones.

All thse facts gave impetus to write this book, "Breeding and Hybridization of Food Fishes". Only food fishes which are cultivated are considered in the book.

K.P. Biswas

Contents

Introduction

The process of manipulating a genetic group of biome in order to change its character to one more beneficial to mankind is called breeding. Breeding practices with regard to fish and other aquatic organisms are so far behind those used in agriculture and animal husbandry as to render comparison inappropriate.

This is because, with regard to marine resources, mankind has thus far continued to concentrate almost completely on the capture of naturally existing resources. Today, however, in Japan as well as most of the rest of the world, aquaculture is accounting for a larger and larger percentage of fishery production. For this reason, it must be said that research in the field of breeding will play an extremely important role in aquatic production in the future.

In breeding, the basic technique involved is selection of superior strains. Subsequently, in order to establish the superior characteristics gained through selection from parent to offspring requires the use of heterosis through the practice of crossing (cross breeding). In this way selection and crossing are two inseperable techniques of the breeding process. Of course, preserving characteristics that occur through mutation is another important part of breeding technology.

The first step in the breeding process is defining its aim. Usually research efforts are directed at achieving superior growth rate, resistance to disease, high prolificacy etc. In some cases, however, like that of Japanese ornamental carp, the object may be body shape or colouring. There are also cases in which the object is to improve meat quality or taste. Therefore, researchers must first of all verify what traits they are looking for and then clarify the way in which those traits are inherited or whether or not they are traits that are not genetically transferred, but rather the products of environment or certain age groups.

Generally, variations in individuals are believed to be the composite effect of a combination of hereditary and environmental factors. However, because the environment is not a constant quantity, it is difficult to judge to what degree a new characteristic is controlled by genetic factors simply by distinguishing between individuals. Advances in breeding technology have, therefore, always followed closely in the foot steps of advances in the science of modern genetics. In other words, breeding research has expanded its realm from the study of individuals to the study of populations and from the study of populations to the study of genetic polymorphism. This has in turn led to a shift in methodology from attempts to create new individuals with certain desirable characteristics, to attempt to bring the contents of the genetic pool of the species in question closer to what mankind would like it to be.

With regard to breeding of fish, it is a well known fact the practice was used in medieval Europe to produce improved strains of carp. This was followed by intense efforts to improve rainbow trout through group selection. Prof. L. R. Donaldson of the University of Washington in the United States began selective breeding for high growth rate trout in 1932 and by 1972 it was reported that a strain had been achieved that reached an average body length of 62 cm as second year fish and 69.1 cm as third year fish. Considering that at the start of the program in 1932 second year fish averaged 36.3 cm and third year fish 46.2 cm, it can be seen that after 40 years a strain with a 1.7 times faster growth rate had been achieved. It must be noted, however, that selection such as this that is based on human priorities, unlike natural selection, proceeds without regard for adaptability to the natural environment of the species. This means that in some cases results can only be achieved under specific conditions. For example, there are cases in which a strain achieved over long years of selection fails to display its acquired characteristics when transplanted to a new environment. Rainbow trout from Donaldson project were transferred to the Nikko Branch of the Aquaculture Research Institute in 1954 and again in 1966. In this case, the first batch of transplanted fish showed a growth rate almost identical to ordinary rainbow trout, while showing a survival rate that was lower than normal.

This result led to their being selected out of the breeding project three years later. A part of the second batch was kept under culture, however, and after 1971 the faster growing portion of the population began to produce three superior strains; one with high prolificacy; one with two spawnings a year and one with larger eggs. Thus, when transplanting strains that are deemed superior, it should be kept in mind that efforts must be made to ensure that they continue to display their superior characteristics in the new environment.

During the century long history of rainbow trout aquaculture in Japan, the biggest results in breeding have come in the area of seasonally earlier spawning. Some 55 years of repeated selection have achieved a spawning period that is three months earlier in the early 1950s as result of the following two desires on the part of aquaculture industry;

1. In order to able to harvest 100 g yearlings in time for the autumn tourist season when seasonal demand reaches its peak, culture operators early eggs spawned in November or December;

2. Operators wanted to realize a production schedule in which groups of trout with different spawning periods could be introduced to the rearing ponds, one after another at 2-3 months intervals, thus enabling continuous year round shipments of mature fish to market.

In recent years, however, this has been achieved more commonly by the use of shade culture techniques to accelerate or delay in spawning.

By the way, although Donaldson strain rainbow trout have not shown significant results in Japanese fresh water aquaculture in terms of growth rate, they are recently being used post smolt stage sea water rearing on an experimental basis in some parts of Hokkaido. Donaldson strain rainbow trout are slow in sexual maturation, taking three years to reach their first spawning. However, this fact means that on the other hand, a period of high growth rate can be expected between the first and second years when the fish are reared in sea water. For this reason, the characteristics of the Donaldson strain are now receiving new attention.

Furthermore since about 1982, a technique of applying a shock to fertilized eggs has led to a non-reproducing strain with triploid genes and subsequently, the start of full scale research into the raising of large size rainbow trout. Concerning the growth of rainbow trout, raising large size fish by means of seawater rearing or biotechnology methods is receiving more attention today than the development of new strains.

However, in these cases too, the parent fish and the fry used must be of superior strains and a thorough knowledge of the strains and the techniques involved with them are still essential.

K.P. Biswas

Chapter 1

Reproduction and Mechanism of Multiplication

The fundamental principle of reproduction is that one or two oranisms give life to a new one. In 1875, it was definitely proved that the basis of the fertilization process in higher organisms lie in the fusion of one female and one male cell (gamets). Pairing of the carrier of hereditary information (chromosomes) takes place in each nucleus and the fusion of these cell nuclei takes place at fertilization. The fertilized egg-cell (zygote) gives rise to a new organism. Sometimes ago, findings concerning certain peculiarities of chromosomes brought to light their role in inheritance. While studying *Echinus* eggs, Boveri (1902-1907) proved experimentally that certain disorders in their development were due to irregular distribution of chromosomes – for normal development the whole set of chromosomes characteristic of the species is required.

It has been established that all the cells of multi-cellular organisms with the exception of generative ones, contain an identical set of chromosomes. As a rule, generative cells (sexual) bear half as many chromosomes as the somatic cells. Somatic cells are reproduced by duplication which is called mitosis or karyokinesis, while generative cells originate by meiotic division.

Mitosis

In proliferating tissues and organs (embryonic tissues, blood forming tissues and so on) the cells are found to be in constant division. The interval between two cell divisions or two mitosis is called interphase. The period covering an interphase and a mitosis is called the mitotic cycle. Interphase and mitosis in their turn are sub-divied into a number of consecutive stages, each playing an important role in the process of cell division. In the course of the interphase the hereditary material

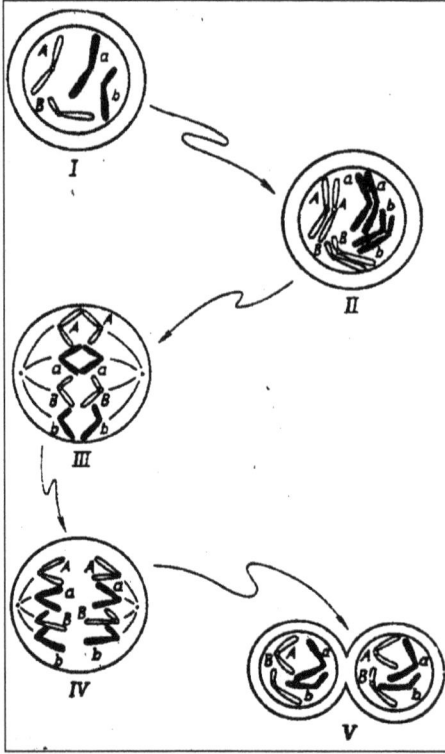

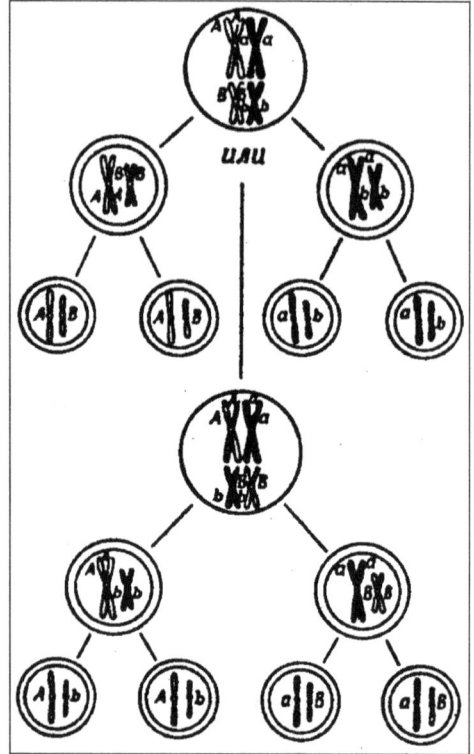

Figure 1.1: Diagram of Mitosis. I: Early Figure 1.2: Diagram of Meiosis.
prophase (chromosomes are observed
in the nucleus); II: Late prophase early
metaphase (chromosomes are doubled);
III: Early anaphase; IV: Late anaphase:
V: Telophase and the end of mitosis
(cytoplasm division).

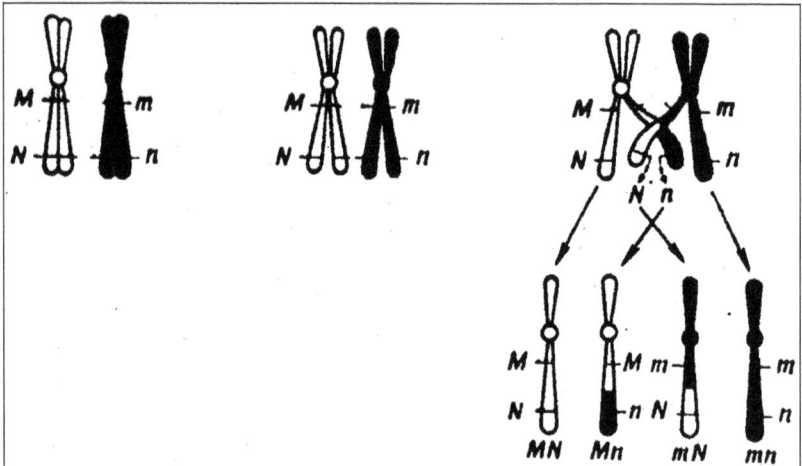

Figure 1.3: Crossover between Chromatids of a Pair of Homologous Chromosomes.

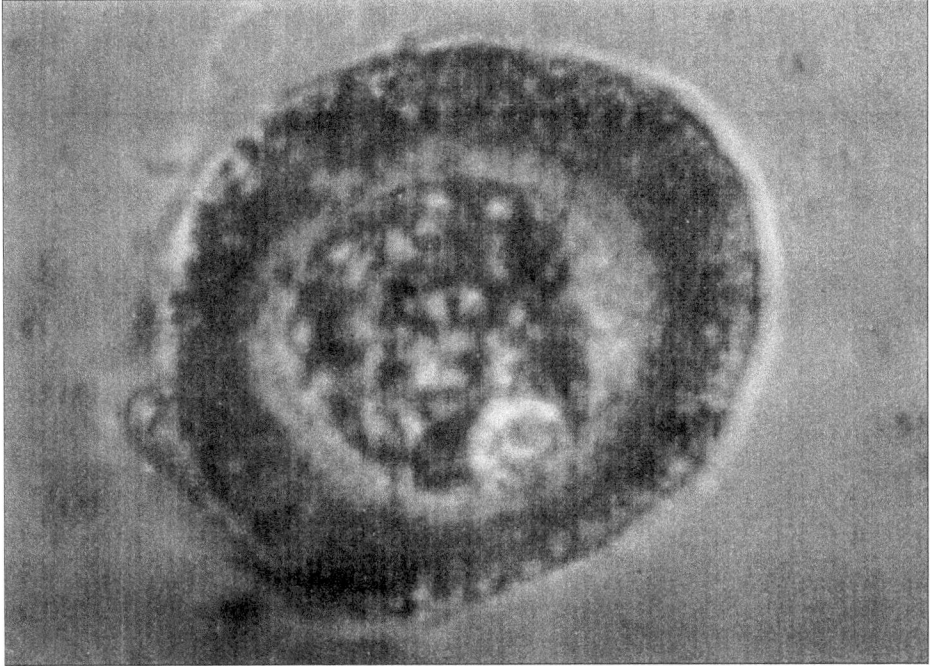

Figure 1.4: Early Prophase I: The Nucleolar Region Appears in Yellow.

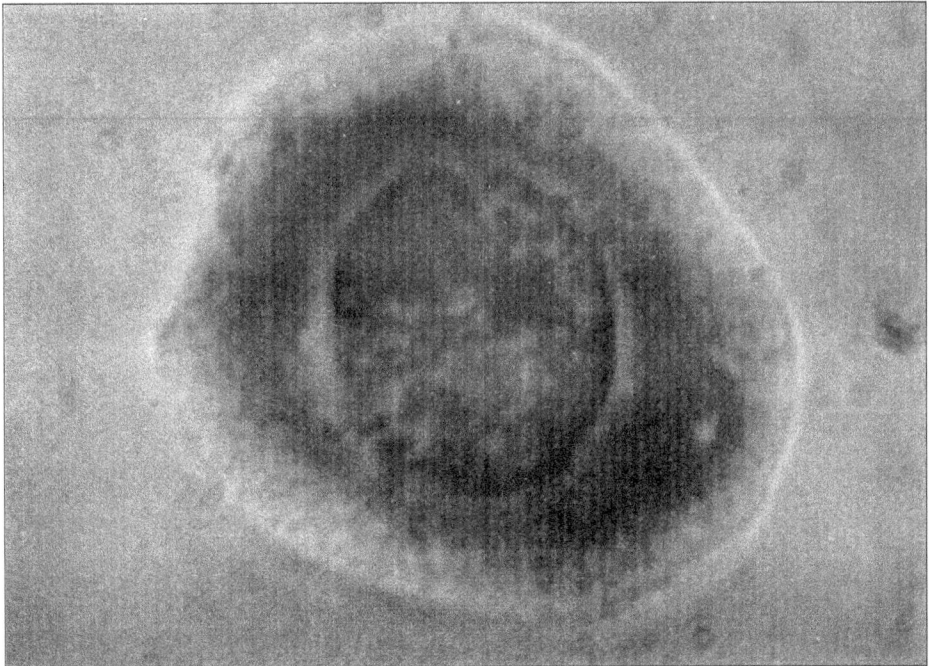

Figure 1.5: Mid-Prophase I. The Denser Nuclear Sphere Appears Dark against the Light Green Cytoplasm.

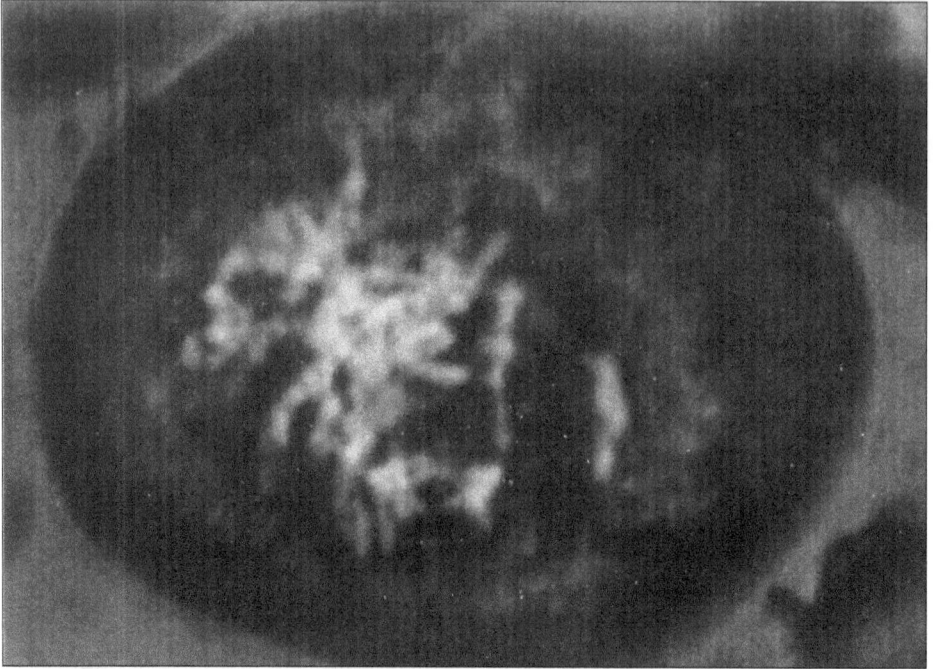

Figure 1.6: Latepachytene: The Nuclear Sphere is Markedly differentiated from the Rest of the Cytoplasm.

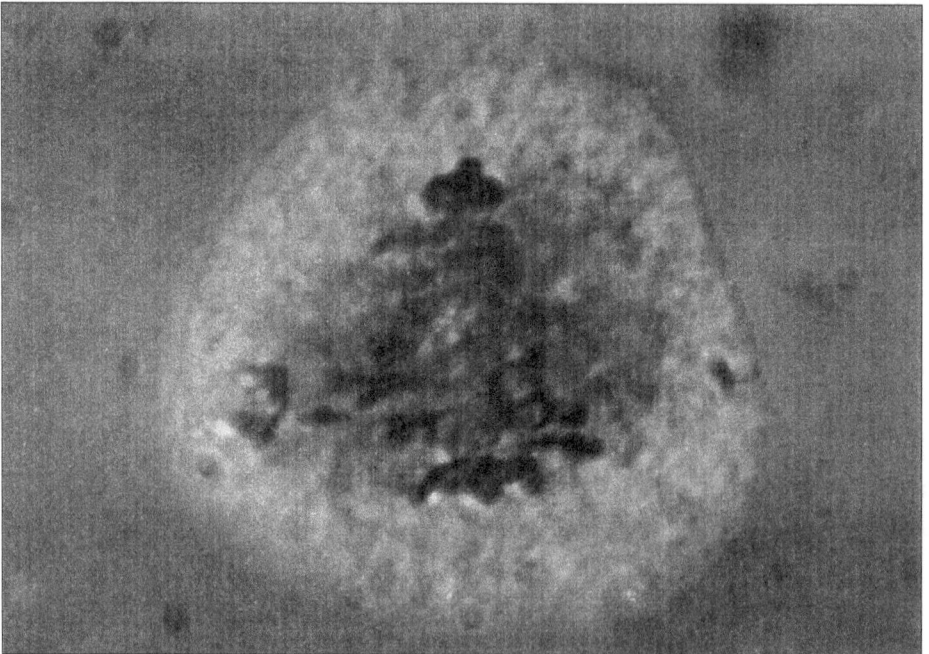

Figure 1.7: Metaphase I: The Chromosomes Appear almost as in an Aceto-carmine-stained Preparation.

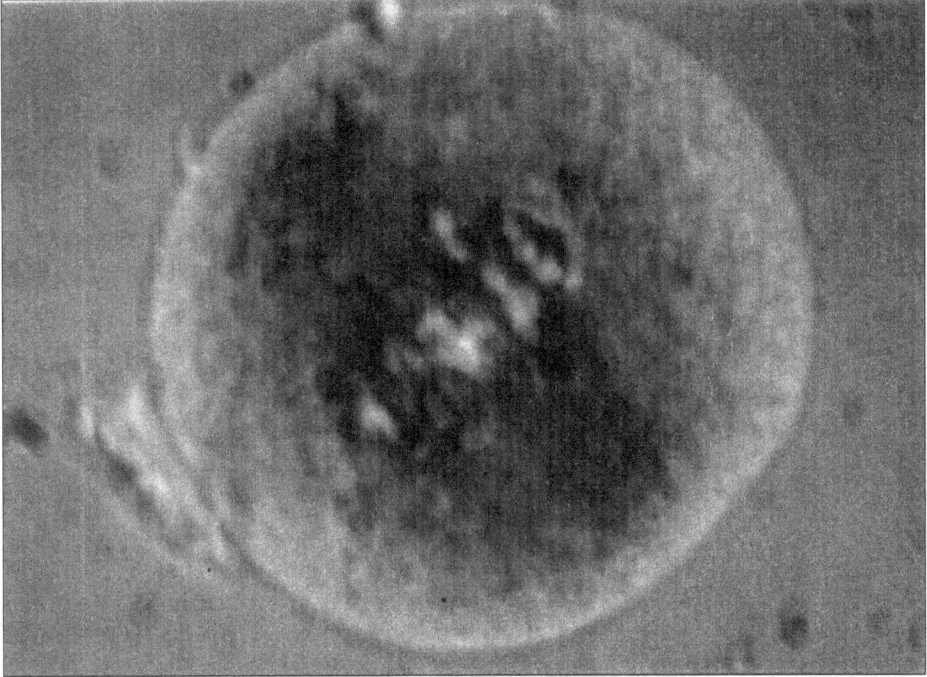

Figure 1.8: Metaphase I: The Spindle Region is differentiated from the Rest of the Cell, Appearing Dark against the Light-blue Cytoplasm.

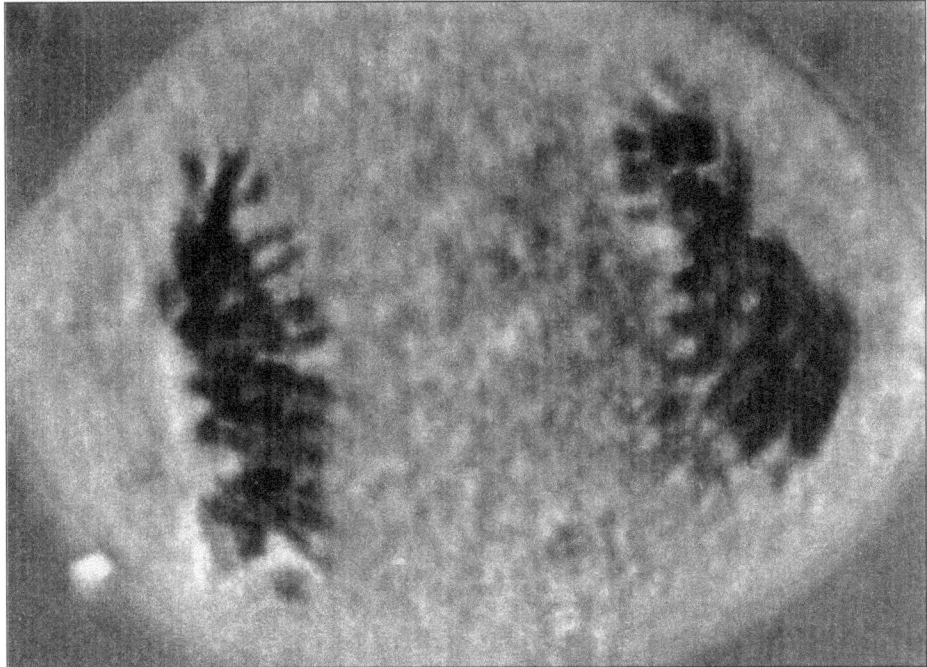

Figure 1.9: Early Telophase I: The Chromosomes Appear almost as in a Stained Preparation

of the chromosome is doubled. The chromosome that had been a single thread structure (consisting of one chromatid) has become a two thread structure. Then at the end of the interphase, just before the mitosis, there occurs a change in the physico-chemical structure of the protoplasm and chromosomes. There appears a protein body, a spindle of division, while the chromosomes are contracting as a result of spiralization. The interphase of the cell transfers into the first stage of mitosis – prophase, at which time the chromosomes can be seen by means of the optical microscope. During the interphase chromosomes are elongated, despiralized, therefore only chromatin net is seen in the interphase nucleus. During mitosis chromosomes coil into a close spiral and at the metaphase-anaphase doubled chromosomes duplicate and each daughter cell gets one chromatid (daughter chromosome). Each chromatid is a replica of the parent chromosome and as genes are localized in the chromosomes, the daughter cell acquire not only karyotype of the parent cell, but it is genotype as well (a set of genes). Mitotic division is the basis of asexual reproduction and is characteristic of certain stages in the life cycle of lower plants and animals.

The process of mitosis is very complicated, but this is justified by the resultant fact that both the daughter cells are identical to the parent cell. There is no loss of hereditary information what so ever as well as no disappearance in a series of cell generations. In the course of mitotic cycle the hereditary information is replicated each time and is transported from cell to cell unchanged.

Meiosis

Each organism develops from the fertilized egg (zygote), which is the result of fusion of the parent gamets. In regular cell division (mitosis) daughter cells get the same set of chromosomes as of the parent cell. In the case of meiosis or reduction division, which results in the formation of sex cells, the number of chromosomes is reduced twice.

Higher organisms are mostly diploid, having two homologous sets of chromosomes in each cell (2n). Cytological findings have proved that the cells forming gamets within a diploid organism experience two consecutive divisions (first and second division) as a result of which the number of chromosomes is reduced twice and gamets having haploid (half) number of chromosomes (1n) are formed. Meiosis in animals producing male sex cells is called spermatogenesis, while the one producing female sex cells is called ovogenesis. Meiosis is the characteristic of all animals and plants that reproduce sexually.

The first meiotic division is characterized by a long prophase during which homologous chromosomes come closer and enter a close elongated conjugation (synapsis stage) forming bivalents or tetraploids. At the time of conjugation, homologous chromosomes, each having two chromatids an exchange their sites, in other words, exchange of hereditary materials between homologous chromosomes take place.

This kind of interchange is called crossing-over and the sites where chromosomes interchange are called chiasmata. The meiosis prophase is transformed into metaphase and then follow anaphase and telophase, contrary to mitosis.

However, at the first meiotic division chromatids of the same chromosome remained paired, while homologous pairs resulting after first division have half the number of chromosomes as compared to the mother cell (2n after the first meiotic division) and the second one starts, in the course of which the cells formed by the first division duplicate mitotically yielding four cells (gamets), each having a haploid set of chromosomes.

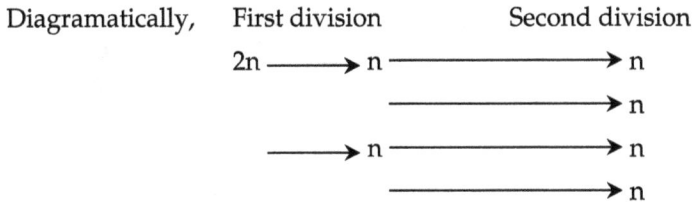

Diagramatically,   First division   Second division

$$2n \longrightarrow n \longrightarrow n$$
$$\longrightarrow n$$
$$\longrightarrow n \longrightarrow n$$
$$\longrightarrow n$$

After spermatogenesis all four cells become gamets, while in ovogenesis only one of the four cells becomes an egg cell, the rest perish.

Another important feature of meiosis is that the duplication of chromosomes in daughter cells occurs in accordance with the theory of probability, that is, by chance. When the chromosomes brought into the zygote by male and female gamets develop into newly formed gamets (during the formation of sex cells in the organism derived from this zygote) they recombine and do not retain their original male and female (parental) characteristics. Therefore, gamets receive both parent's chromosomes in equal number. Pairing (conjugation) and separation of homologous chromosomes is responsible for the mechanism of segregation of allele genes. Independent assortment of genes occurs as a result of an occasional orientation of each chromosome to one or another pole during meiosis providing for purity of gamets. Thus daughter cells during meiosis (unlike in mitosis) receive a different set of genes, that is, they have different genotypes.

Double reduction in the number of chromosomes when gamets are derived is due to the fact that in the process of fertilization there occurs chromosome pairing of fusing gamets. If this pairing is not compensated by double reduction of chromosomes in gamet formation, each subsequent generation would have double the number of chromosomes, that would lead eventually to an unlimited increase of chromosomes within several generations. Double reduction of chromosomes at meiosis ensures a constant nmber of chromosomes in organisms.

The Chromosome Theory of Inheritance

The chromosome theory of inheritance that was completely developd in the second decade of the 20th century by the American school of genetics headed by T. H. Morgan managed to explain the phenomenon of linkage. This theory did not only explain and connect all the facts of the so-called exceptions from Mendel's laws, but it also appeared to be a sound basis for the whole structure of modern genetics.

To explain the phenomenon of coupling, Morgan put forward a suggestion that only those genes that are in the same chromosome are inherited together. Further experiments helped to established the fact that there are groups of genes (linkage groups) tending to be inherited together and that the number of these groups of

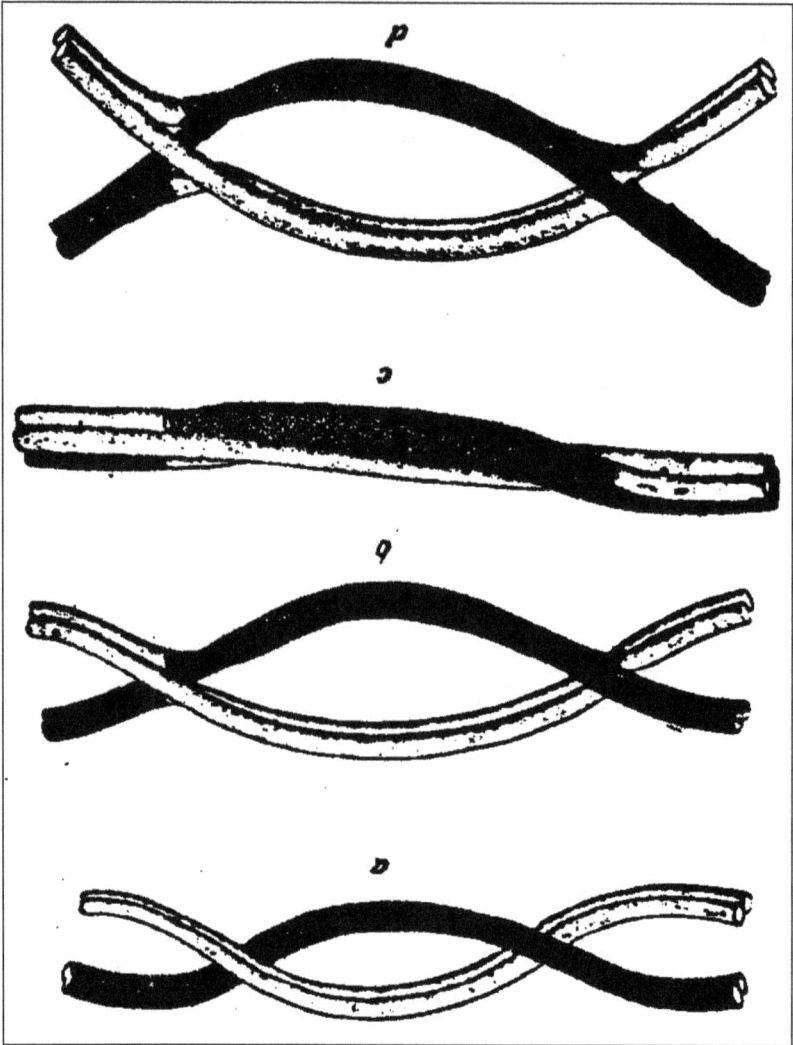

Figure 1.10: Diagram Illustrating Crossing-Over Effect between Two Anterioposteriorly Segregated Chromosomes.

genes never exceeds the number of pairs of homologous chromosomes in a given organism, that is, it is equal to the haploid set of chromosomes. Thus, the idea of linkage groups is the first point of the chromosome theory of inheritance.

The next point of the chromosome theory of inheritance concerns the linear order of genes in the chromosomes. Figuratively speaking, a chromosome is like a thread of beads where every bead is a gene. A thorough study of the degree of different gene linkage within the same chromosome made it possible to determine the position of genes within one chromosome, as well as their sequence in relation to each other.

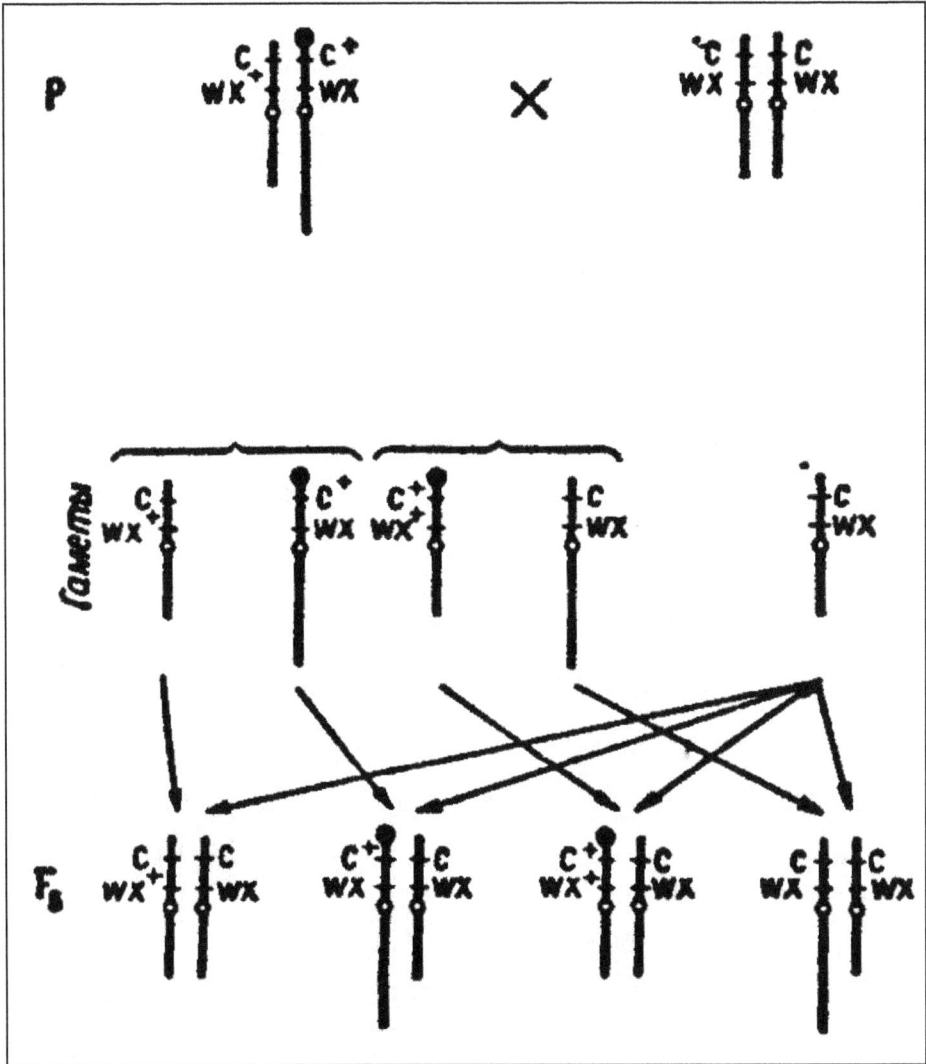

Figure 1.11: Cytological Proof of the Crossing-over in *Zea mays*.
C⁺: Stained endosperm; c: Unstained; wx⁺: Starchy; wx: Waxy

Experiments showed that the degree of linkage between two given genes was always the same- the frequency of recombinations between them appeared to be always the same, but for the different pairs of genes the frequencies were different. To explain this phenomenon Morgam made a supposition that genes found in one chromosome close to each other are tightly linked. Linkage of genes placed at some distance from each other is less manifest. In other words, he believed that the degree of linkage is inversely proportional to the distance between the genes.

Crossing-over is another hypothesis put forward by Morgan to explain the linkage failure. According to this hypothesis there takes place an exchange of homologous sites between the chromosomes of homologous pairs in bivalents during the first meiotic division. The break up in gene linkage occurs due to the break up of chromosomes at the point between the given loci (a locus is the site of gene location in the chromosome which is always the same for the given gene). It is also due to the recombination of chromosome sites occurring there. In the course of the process the site of one chromosome contacts with the site of the other. Two new chromosomes are derived, each having only one of the two genes previously found in the same chromosome. A number of crossing-over at different sites may occur between one pair of homologs; in case of crossing-over, however, the possibility of the second crossing-over close to this site is limited (such phenomenon is called interference). Thus, the third point of the chromosome theory of inheritance is crossing-over, the mechanism leading to the exchange of genes or homologous sites between chromatids of homologous chromosomes at meiosis.

Crossing-over: Genetic Mapping

The studies carried out by C. Stern (1929) and McClintock both cytologically and genetically showed that assortment of linked genes is accompanied by a site exchange of homologous chromosomes. Independent assortment of genes and splitting in hybrid progeny according to Mendel is secured by chromosome behaviour at meiosis, independent assortment being possible only for those genes that are located in non-homologous chromosomes.The mechanism of meiotic assortment (reassortment) is related to the behaviour of homologous chromosomes during meiosis.

Crossing-over has ceased to be a hypothesis and has become a practically confirmed phenomenon. The period of a site exchange between chromatids at crossing-over is now associated with the time of chiasma appearance at meiosis. It is now considered that chiasma occurs where exchange of sites takes place. The mechanism of crossing-over is still obscure.

Chromosome Mapping

The study of crossing-over brought about the idea of a linar order of genes in a chromosomes, which in turn made it possible to have genetic maps of chromosomes showing interposition and distance between the genes of a given linkage group.

According to Morgan's supposition the distance between the genes in a chromosome is proportional to the number of crossings between them. This served as a basis for the determination of gene position inside the chromosome.

If in two genes A and B crossing-over is rare, say 50 percent of the cases, it can be supposed that they are located rather close to each other in the chromosome which results in a small probability of a chromosome break up at the site and consequently of the site exchange. Thus, if the number of gene exchanges is proportional to the distance between them, the measurement of the crossing-over frequency between A and B as well as between B and C (Figure 1.12) is to indicate the distance between the genes A and C, because the distance between A and B plus the distance between B and C is to be equal to the distance AC (the distance ubit between the genes was considered to be the chromosome site where 1 percent of crossing-over occurs).

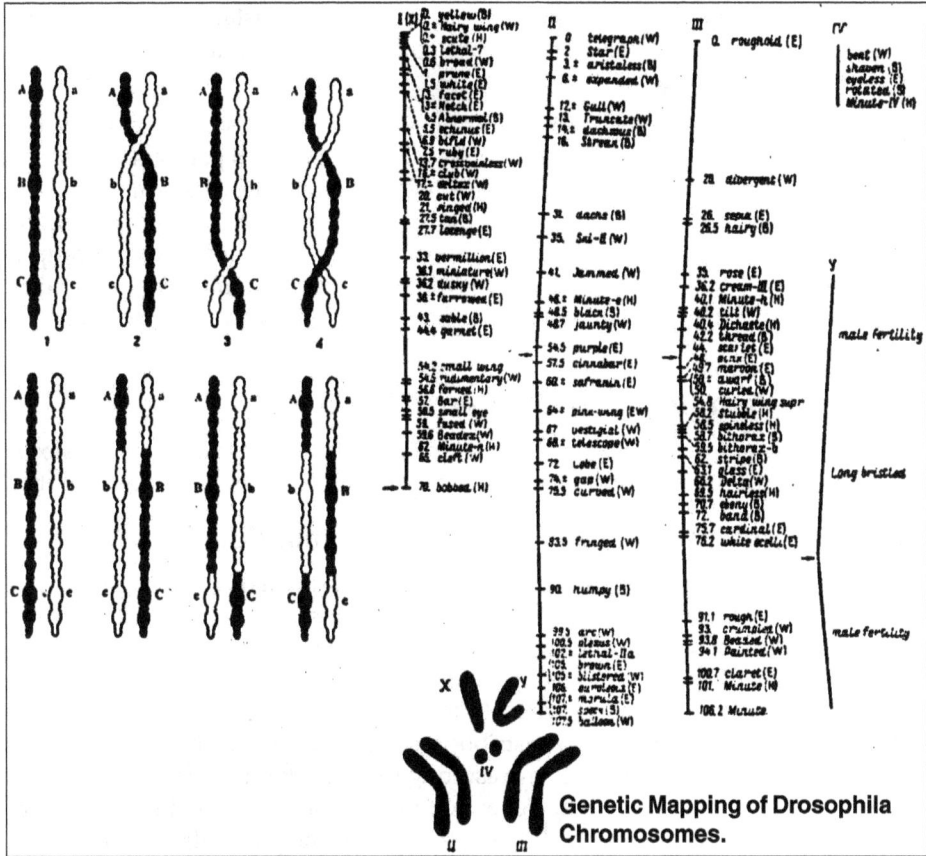

Genetic Mapping of Drosophila Chromosomes.

Figure 1.12: Diagram of the Mechanism and Crossing-over Effect in a Pair of Homologous Chromosomes.

1: Lack of crossing-over; 2: Crossing-over between A and B; 3: Crossing-over between B and C; 4: Double crossing-over between A and B, B and C.

Most cases prove occurrence of such dependence. Thus, in crossing of red-eyed, yellow-bodied *Drosophila* with white-eyed grey-bodied forms, the crossing-over between the yellow-body gene (yellow) and the gene of white eyes (white) can be observed in 1.5 percent of the cases. The crossings between white-eyed flies with normal wings and red-eyed flies with bijid wings (bijid) cause the crossing-over between the gene of white eyes and bijid wings in 5.5 percent of the cases. If the gene of white eyes lies between the genes of yellow body and bijid wings, it can be expected that the distance between W and Y (1.5) plus the distance between Y and b (5.5) should be equal to the distance Y and b (7.0).

It should also be noted that since the genes occupy a definite position inside the chromosomes the percentage of their crossing-over is always the same. An analyzing crossing, that is, crossing with a recessive homozygote, is nearly always used to determine the frequency of crossing-over. As known, an analyzing crossing helps to detect all types of gamets that are found in the analyzed hybrid species.

On genetic maps, however, one can often come across distances between genes of 100 units or more. These distances are calculated by summing the distances between intermediate genes.

When making a genetic map, a linkage group is the first to be determined; that is, the largest possible number of genes found in the given pair of homologous chromosomes. Then by means of crossing-over, a relative interposition of genes in the linkage group is determined. In genetic mapping a definite system of designation of genes accepted for such organism is used. In such maps the linkage group is sure to be shown as well as the distance from one of the chromosome ends taken as a zero point, the former being expressed in morganida, as well as the site of the centromere.

Determination of Sex and other Related Matter

The most common type of sex determination is syngamic, when sex is determined at the moment of gamet fusion in fertilization (mammals, birds, fish etc.).

Chromosome Mechanism of Sex Determination

Sex as well as any other character of the organism is hereditarily determined. The most important role in genetic determination of sex and the maintenance of equal correlation of sex is played by the chromosome mechanism. In other words, sex is determined in the process of fertilization by the chromosome set of the fusing gametes. Most animals give birth to an equal number of male and female individuals, which means that sex distribution is close to 1:1, which can be observed when analyzing crossing. In progeny resulting from this kind of crossing the genes are split in the ratio of 1Aa : 1aa. Genes A and a should be located in one pair of chromosome. If sex is inherited following the same principle as other characters, it may be supposed that one sex is homozygote while the other is heterozygote. Then the sex distribution in the progeny will be 1:1.

The above supposition was first made by Mendel. Later on genetic investigations as well as cytological studies made in early 20th century, proved the existence of the chromosome sex-determining mechanism for most animals and a number of plants. Females of many animals were found to have all chromosomes paired and to produce only one kind of gamets at gametogenesis. Male produce two different kinds of gamets, one similar to those found in the female and another differing in structure of one of the chromosomes. A homologous pair of chromosomes (determined as such by their behaviour at meiosis) but dissimilar in size and shape, appeared to be connected with sex determination. Such chromosomes have been called sex chromosomes. It has been shown that the fusion of an egg-cell carrying a sex chromosome (X-chromosome) with a sperm carrying the same chromosome will give rise to a XX zygote and to the development of a female individual. The fusion of an egg-cell with the sperm carrying another sex chromosome (Y-chromosome) will give rise to an XY zygote and to the development of a male.

Furthr investigations revealed different ways of chromosome sex determination. It appeared that homogamy (homogametic sex is the one producing only one kind of gamets, heterogametic sex producing two kinds) may be the attribute of not only the female sex, but also of the male – some fish species, carp for example probably have homogametic sex determination.

There are also a number of modifications of the modes of sex determination described above, depending on the balance of sex chromosomes and autosomes in the zygote. All these methods of chromosomal sex determination have one thing in common, that is, the sex ratio in the progeny is conditioned by a random combination of homogametic and heterogametic sex gamets that results in the primary zygote yield of male and female types in the ratio of 1:1.

Sex-Linked Inheritance

The genes located in sex chromosomes are called sex-linked and the inheritance of such genes (and hence their related characters) is called sex-linked inheritance. Which sex brings about dominant and recessive characters, has no relation to the splitting according to the given characters in hybrid offsprings. This is true when genes are located in autosomes equally reflected in both sexes. When genes are located in sex chromosomes the nature of inheritance and splitting depends on the chromosomic behaviour at meiosis and then combination in fertilization. Genetic investigations of a number of subjects (*Drosophila*, main) showed that chromosome of heterogametic sex does not carry any gene, that is, it is inert in respect to inheritance. These follows an important practical conclusion: recessive genes of X chromosomes are developed in heterogametic sex, since no dominant alleles in X chromosome are opposed to them.

As sex-linked characters are of great practical importance, it may be observed that genes located in sex chromosomes are inherited criss-cross. Sex chromosome of homogametic organisms are passed on to both sons and daughters, while only the X chromosome of heterogametic sex is handed on either to daughter (in the case of male heterogamy) or to sons (in female heterogamy). If there is a definite tendency of crossing and the characters depending on X chromosome are handed on from mother to sons and from father to daughters, then the inheritance is called criss-cross. In sex-linked characters the appearance hybrid F1 depends on which character has been contributed by the father and which one by the mother. In some lines of crossing, one come across apparent contradiction to Mendel's first law of dominance or uniformity of progeny F1.

An important conclusion based on the study of the sex-linked characters is that these characters may serve as markers which can help to distinguish animal sex as early as in F1.

Sex-linked or sex-depending characters are those expressed exclusively or mainly in one sex. Secondary sex characters in men and animals, as well as a number of other characters may be reffered to as such. Manifestation of these characters is apparently connected with the effect of male and female hormones.

Sex Reversal in Ontogenesis

The effect of certain genes as well as hormones in the process of the organism's development may result in sex reversal is known to exist in plants, *Drosophila*

(Andman). Hormonal sex reversal, however, is of great practical value for cattle breeding, a means of artificial regulation of sex ratio. The following experiment made by T. Yamamoto in the 1950s is of interest to fish breeders.

Yamamoto experimented on aquarium fish, *Oryzias latipes* whose heterogametic sex is male. White as well as red *Oryzias* occurred in the experiment, the red gene being carried by Y chromosome, while its recessive allele r is carried by X chromosome. The males were red (Xr YR); the females were black (Xr Xr). In this case males were always red, since they carried the dominant allele, R. The sons of this type of inheritance will always have the father's character (if no crossing over occurs between X and Y, which is very rare). Crossing of Xr Xr and XrYR always resulted in white females and red males. Hatched young fish, even before sex differentiation, were divided into two groups. One group had a routine food ration, the other received female sex hormone (estrone and stilbestrol) added to the food ration. As a result, all red fish in the second group determined genotypically, as male Xr YR (red), turned out to be females according to phenotype, having normal ovaries and female secondary characters. They were able to cross with normal red males. The crossing of such females with normal males (Xr YR X xr yR) resulted in the sex ratio of 1 (Xr Xr):30 (2Xr YR and 1 YR YR), not 1:1.

Yamamoto's investigation clearly demonstrates the possibility of sex differentiation in ontogenesis.

Thus the study of sex determination mechanisms and the behaviour of sex-linked characters enables to conclude that;

(a) Sex of the organism is inherited as many other character determined by a gene;

(b) The sex ratio 1:1 results from the formation of two kinds of gametes having equal frequency in heterogametic sex at meiosis;

(c) Heterogametic sex may be both male and female;

(d) Inheritance of the sex-linked characters is predetermined by the genes located in sex chromosomes.

Chapter 2

Natural Breeding of Fin Fishes

Physiology of Breeding Mechanism

Reproduction in fishes is dependent on the coordinated actions of various hormones associated with the brain-hypothalamus-pituitary gland axis.In general, ovarian and testicular function in teleosts and other fishes is controlled not only by the pituitary gonadotropins but also by multiple hormones and growth factors which act in an endocrine, autocrine or paracrine manner. The activities of gonadotropins are under the control of hypothalamic neuroendocrine factors, released at the level of pituitary, paracrine factors from within the pituitary, and through the feedback cations of gonadal hormones. In turn, the secretion of hypothalamic factors is regulated directly or indirectly by sensory inputs, feedback actions of the pituitary-gonadal hormones and other hormones. Thus, at each level of this axis, a limited number of target cells are under the influence of many factors. The final cellular response results from integrated effects of these regulatory factors on the intracellular signal transduction components.

Pituitary Gland and Hypothalamus

The pituitary (hypophysis) is an endocrine gland located on the ventral side of mid brain and is attached to it by means of stalk. This has two components, that is, neurohypophysis and adenohypophysis. Neurohypophysis is derived from downward evagination of third ventricle. Adenohypophysis is derived from the embryonic pouch, Rathke's pouch, arising from the roof of buccal cavity as an outward protrusion. The neurohypophysis consists of nerve fibers, arising from perikaryon of neurons, which are located in hypothalamus. The cluster of neurosecretory neuron is called preoptic nucleus or nucleus preopticus (NPO). These neurons secrete nannopeptides, such as agrinine, vasotocin and isotocin. Isotocin is similar to vasopressin and oxytocin of higher vertebrates. The neurosecretory

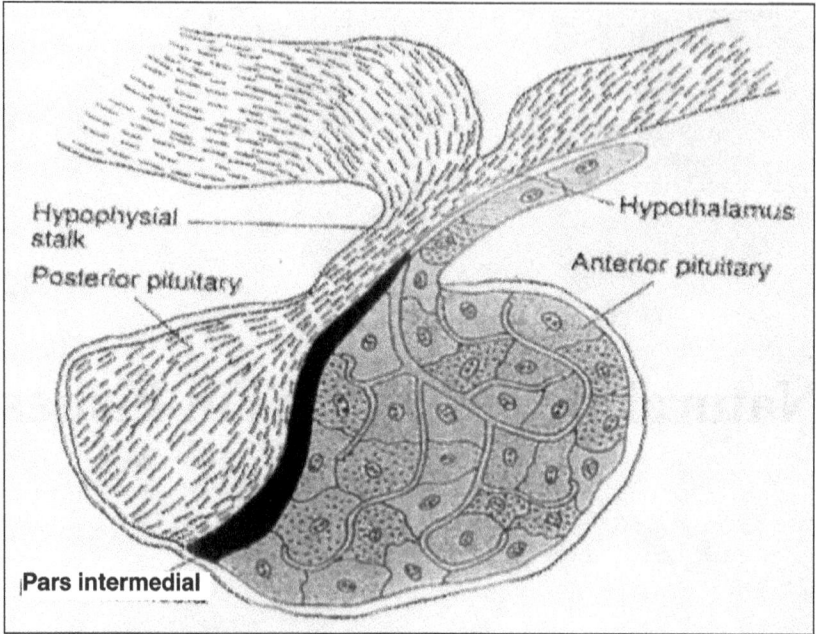

Figure 2.1: Pituitary Gland.

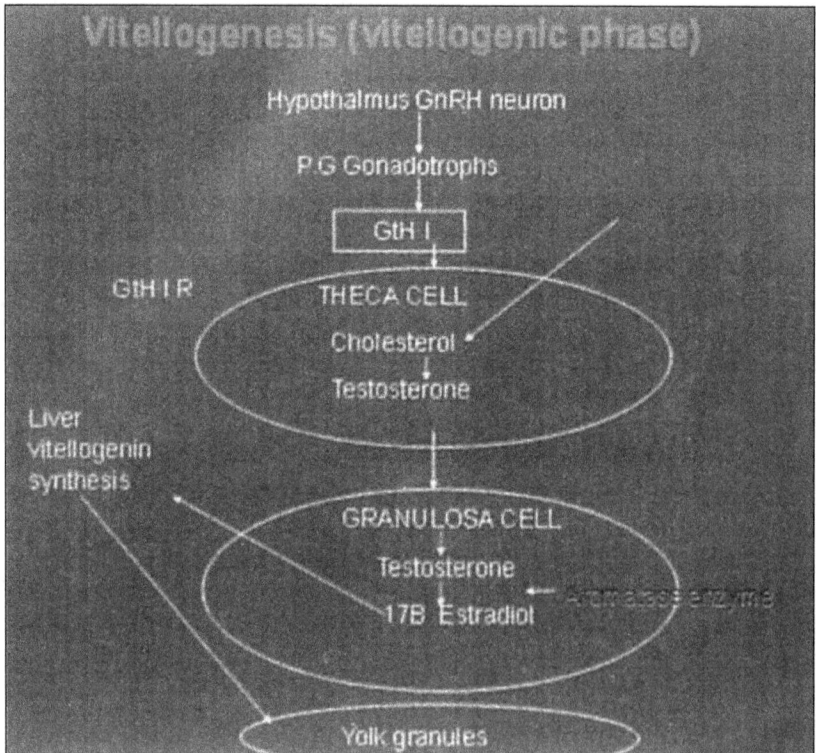

Figure 2.2: Showing Role of GtH I in Vitellogenesis.

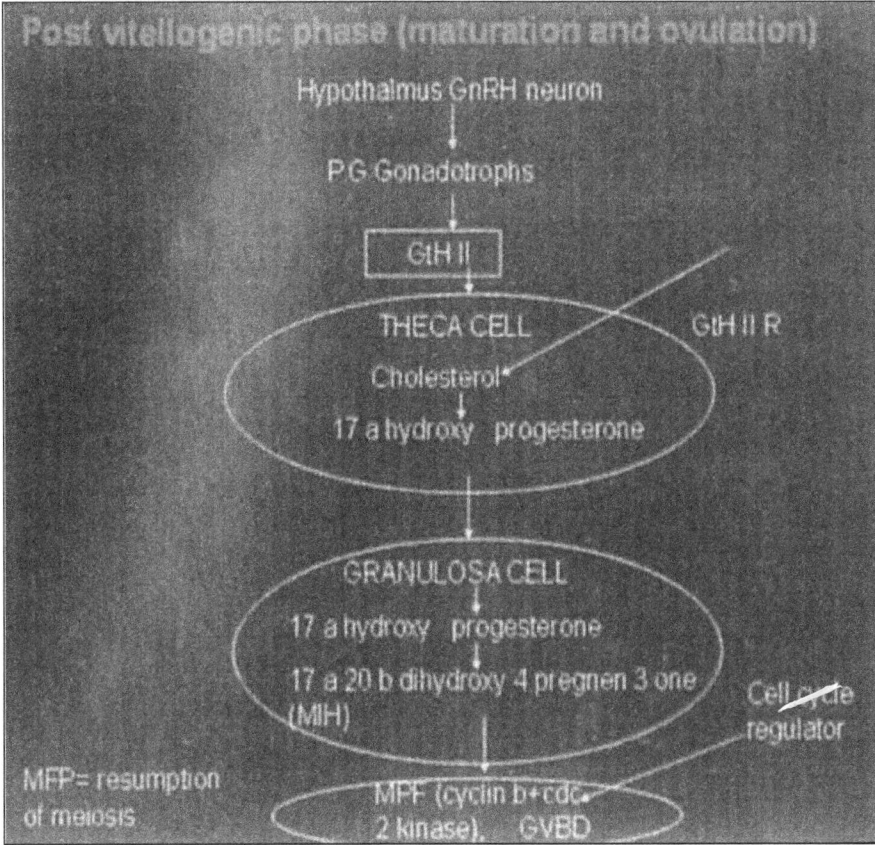

Figure 2.3: Showing Role of GtH II in Maturation and Ovulation (cdc: Cell division cycle).

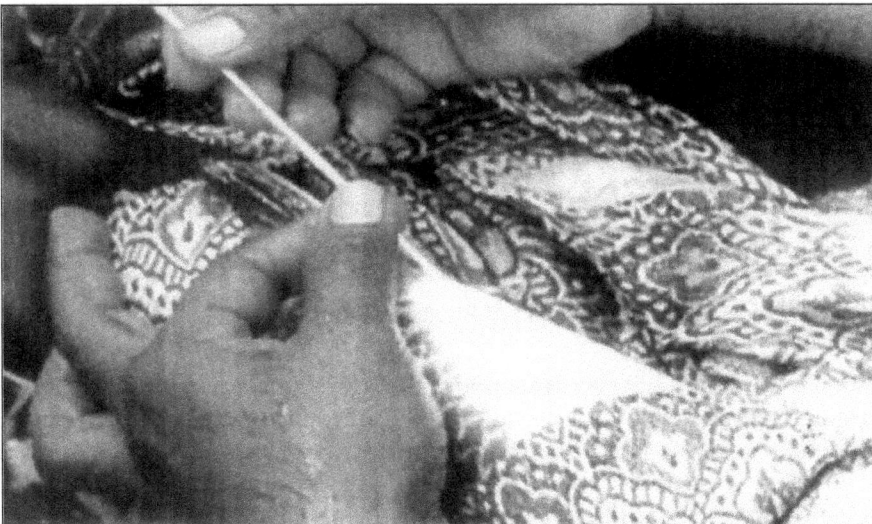

Figure 2.4: Cannulation to Assess the Maturity.

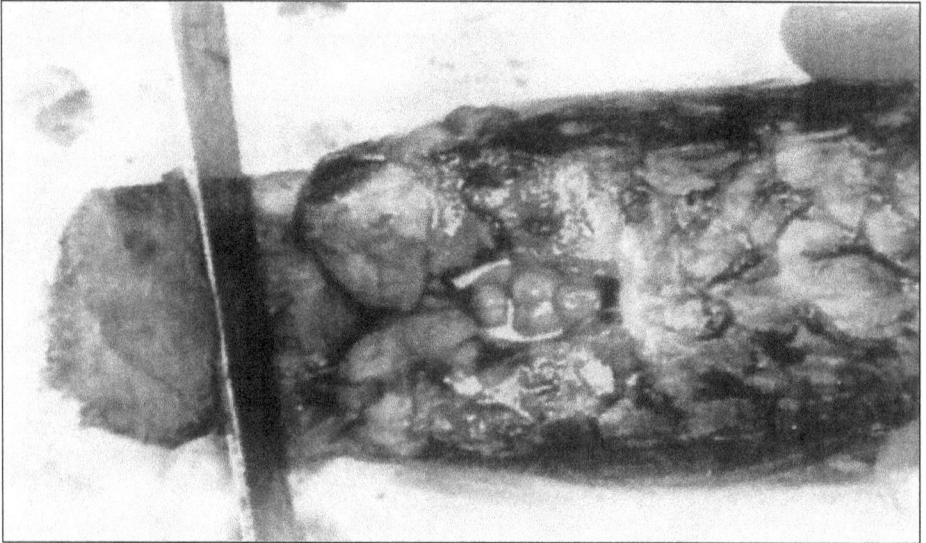

Figure 2.5: Small Arrow Showing Position of Brain and Large Arrow Pituitary Gland under the Brain.

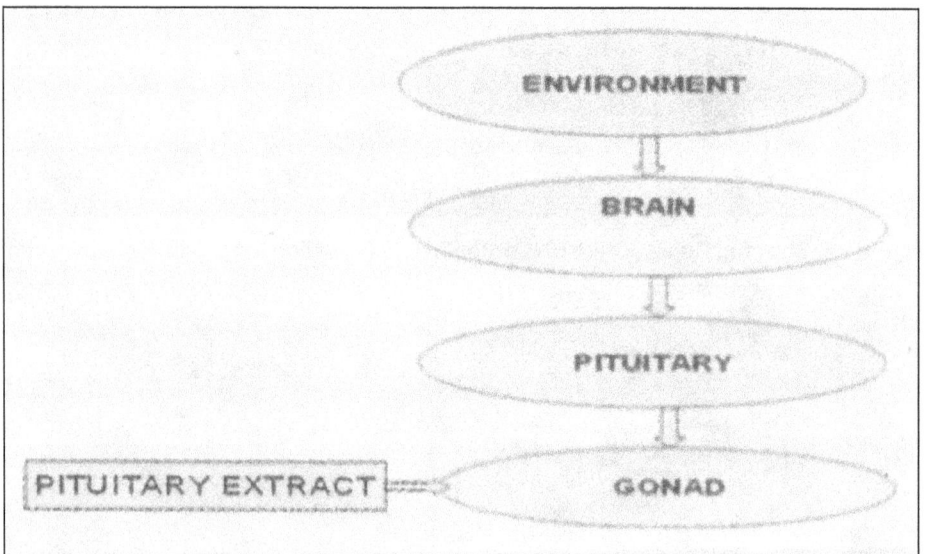

Figure 2.6: Showing Mode of Action of Pituitary Gland Extract.

hormones from the neurons of preoptic nucleus are transported through the axon and stored in the axon terminals ending at the junction between neurohypophysis and adenohypophysis which is called neuroadeno interface. This interface is highly vascular and it is the center for storage and release of neurohormones. Hence it is called neurohaemal area. Apart from NPO, there are several other clusters of neurosecretory neurons in the hypothalamus out of which nucleus lateralis

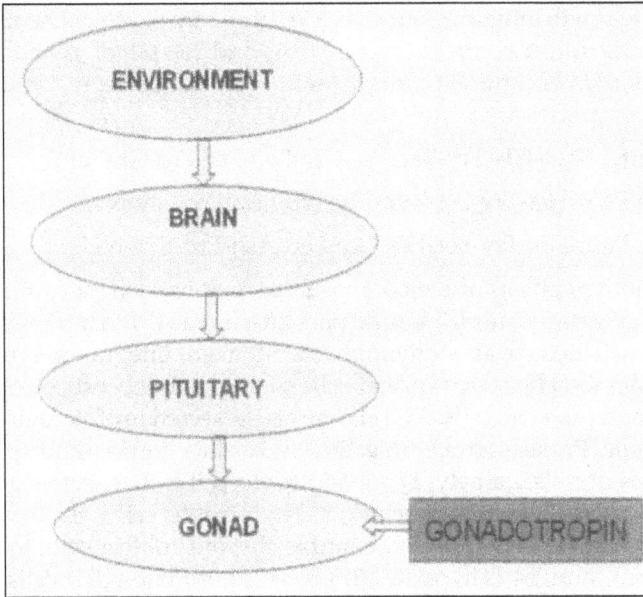

Figure 2.7: Showing Mode of Action of Gonadotropin.

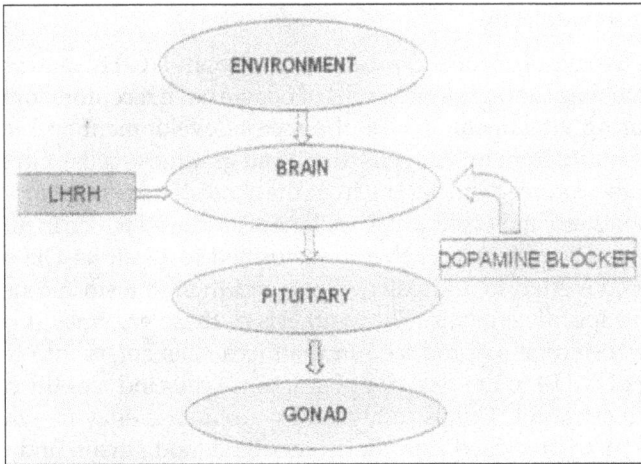

Figure 2.8: Showing Mode of Action of LHRHa and Dopamine Blocker

tuberis (NLT) is extensively studied. In the case of teleosts these neurons also send their axonal projections to neuroadeno interface. The secretion of these neurons controls the adenohypophysial cells. Hence, they are called adenohypophysiotropic neurons or neurons secreting/releasing hormones or inhibiting hormones. These hormones stimulate the release of adenohypophysial hormones and inhibiting factors (hormones) inhibit the secretion of adenohypophysial cells. Three adenohypophysiotropic hormones isolated so far fron hypothalamus are GnRH (Gonadotropin releasing hormone), TRH (Thyroid releasing hormone) and somatostatin (Growth inhibiting hormone). All three are peptides chemically. GnRH is a decapeptide and it controls the synthesis and release of gonadotropin from gonadotrophs cells of pituitary gland. Amino acid sequence of GnRH molecules in different fish species are as follows;

Sea Bream : Pglu-His-Try-Ser-Tyr-Gly-Leu-Ser-Pro-Gly-NH2

Salmon : Pglu-His-Try-Ser-Tyr-Gly-Trp-Leu-Pro-Gly-NH2

Cat fish : Pglu-His-Try-Ser-His-Gly-Leu-Asn-Pro-Gly-NH2

The adenohypophysis of a teleost has three regions such as rostral pars distalis (RPS), proximal pars distalis (PPD) and pars intermedia (PI). The neurohypophysis integrates into all areas of adenohypophysis but major integration is found in pars intermedia. The RPD has two types of cells such as Adrenocorticotropic hormone (ACTH) cells and prolactin cells. ACTH controls the secretion of steroidogenic cells of interregnal gland. Prolactin from prolactin cells has function in osmoregulation. PPD has three types of cells, namely, 1) gonadotrophs cell is that secrete gonadotropin (GtH) which controls gonadal function, 2) Thyrotrophs cell is that secrete thyroid stimulating hormone (TSH) which controls thyroid follicle cells to produce T3 (Triidothyronin) and T4 (Thyroxin), 3) Growth hormone (GH) cells that secrete growth hormone and which controls the normal growth. The PI has two types of cells, namely, Melanocytes stimulating hormone (MSH) cells and PAS somatolactin cells secreting somatolactin.

There are two types of gonadotropin; gonadotropin-I (GtH-I) and gonadotropin-II (GtH-II). Both theca and granulose cells of oocyte have receptors for GtH-I. GtH-I is secreted during vitellogenic phase of oocytes development and it induces the secretion of estradiol from follicle cells (theca and granulose cells). GtH-I stimulates the theca cells to produce testosterone from the cholesterol precursor. By side chain cleavage of cholesterol molecule, pregnenolone is produced which in turn is changed to progesterone. Progesterone is then transformed to 17 alpha-OH progesterone which is then converted to androstenedione and then to testosterone. Each stage is catalyzed by specific enzyme. The synthesis of these enzymes is controlled by GtH-I. The testosterone so produced in the theca cells enters into the granulose cells. Binding of GtH-I to the receptor of granulose cells induces the conversion of testosterone to estradiol. This is catalyzed by aromatase enzyme which is under control of GtH-I. Synthesized estradiol enters the blood stream and reaches liver cells to produce vitellogenin, which is precursor of yolk protein. Vitellogenin reaches oocyte cytoplasm through blood and get transformed into yolk protein (lipovitellin and phosvitin) and this gets deposited as yolk granules in ooplasm. GtH-II stimulates the follicle cells of oocytes to produce Maturation Inducing Hormone (MIH) which is 17 alpha-20-beta dihygroxy-4-pregnene-3-ones (17 alpha-20-beta dihydroxy progesterone) from 17 alpha-OH progesterone. Thus, there is shift in steroidogenesis at this stage. MIH stimulates the production of maturation promoting factor (MPF) in the cytoplasm of oocytes. MPF is formed by the union of molecule of cyclin B and cdc 2 kinase (Cell division cycle 2 kinase). MPF brings about the following changes during final maturation of oocytes;

1. Migration of germinal vesicle to the animal pole of the oocyte (towards micropile);
2. Germinal vesicle breakdown (GVBD) takes place;
3. Condensation of chromosomes and alignment in the metaphase spindle;
4. Division of nucleus and extrusion of first polar body;
5. Coelescence of yolk globules and lipid droplets and increase in the transluscency;
6. Increase in oocyte volume/size.

Two chemically different gonadotropins, GTH-I and GTH-II, have been characterized in species representing four orders of teleosts. GTH-I and GTH-II contain a common alpha subunit together with a distinct beta subunit and are generally conceived to be FSH-and LH-like respectively. GTH-I and GTH-II have distinct roles in controlling gonadal development. GTH-I is generally believed to be important for vitellogenesis and early gonadal development, while GTH-II stimulates events leading to final oocyte maturation and ovulation in females and spermiation in males. Gonadotropins play a pivotal role in controlling reproductive development in teleosts.

The hypothalamic control of GTH-I and GTH-II release in teleosts is affected by the seasonal reproductive cycle or gonadal maturation cycle of the animal. In salmonids, the effectiveness of gonadotropin releasing hormone (GnRH) to increase GTH-I and GTH-II is favoured towards GTH-I secretion during early and late stages of vitellogenesis in females and during spermatogenesis in males. As gonadal maturation advances beyond vitellogenesis in female and spermation begins in males, the GTH-I response decreases relative to GTH-II.

Environmental cues, such as photoperiod, are known to affect reproduction in teleosts. Photoperiod affects gonadal growth and maturation, the daily cycle of plasma GTH-II levels and the GTH-II release response to GnRH analogues.

Other external factors, such as pheromones also affect GTH-II secretion. In gold fish, exposure to steroid hormonal pheromone (released by females during the preovulatory period) increased neuronal activity in the olfactory nerve and elevated GTH-II release in males.

With the diversity of reproductive styles (oviparity, ovoviviparity, viviparity) fish exhibits- perhaps the widest array of reproductive strategies among the vertebrates. While progress has been made in understanding the hormonal factors controlling gonadal growth, differentiation and regression relatively little is known concerning the hormonal mechanisms regulating gametogenesis in comparison with mammals.

Experiments using unilateral gonadectomy illustrate the active role of cell proliferation in fish. Two gonadotropins (GTH-I and GTH-II) have distinct roles in all teleosts.

Although most studies have focused on aspects of cell proliferation, growth and differentiation, cell death may be an important factor involved in all aspects of gonadal physiology. Development of a cohort of mature oocytes and sperms requires a mechanism to select only viable gamets. This may be particularly relevant in female fish where large number of developing ovarian follicles are recruited into the growing oocyte pool. In seasonally breeding species cell death processes may also be involved in the regression of gonadal tissues. A variety of environmental stresses, such as, hypoxia, temperature or inadequate nutrition may result in the regression of gonadal tissues. In addition anthropogenic stressors, such as, environmental toxins may also compromise gonadal development by increasing the rate of cell death. Recent studies suggest that anthropogenic factors which affect reproductive development in fish may also do so in part through effects on apoptosis. For example, white sucker populations down stream of kraft pulp mills in Ontario consistently display evidence of altered plasma sex steroid hormone concentrations, reduced gonad size, decreased fecundity and delayed sex maturation. Certain components of bleached pulp mill effluent act to stimulate apoptotic cell death either directly or indirectly via alterations in endocrine homeostasis.

Hormonal and Environmental Manipulation of Maturation in fish

Most teleosts are seasonal spawners and their maturation and spawning normally coincide with optimum environmental conditions. Among the different environmental factors involved in reproduction events, photoperiod (duration of light) and temperature have been found to be of prime importance in most fishes. It is known that external environmental stimuli or triggers are received by exterorecepters (eyes and pineal glands are known to provide information about day/night, season etc.)which transfer this message to the brain of fish. The hypothalamus (an endocrine gland present in the base of the brain) releases a chemical messenger known as gonadotropin releasing hormone (GnRH) which then acts on the pituitary to release gonadotropin. The gonadotropins stimulate gonads to synthesize steroid hormones which regulate gonadal growth, secondary sexual characteristics, maturation and ovulation or spermiation in fish.

In female fish, the reproduction process involves three basic steps namely, maturation, ovulation and spawning. Maturation is a process by which tiny sex cells (germ cells) develop, along with accessory tissues, develop into large gonads (ovary and testis) which finally lead to the production of gamets (eggs and sperms). Females are considered mature when final maturation occurs. Final maturation is indicated by migration of nucleus to the periphery of oocytes and breakdown of germinal vesicle. Ovulation is the release of mature oocyte from the ovarian follicle in to the lumen/body cavity. Spawning refers to the expulsion of oocyte to the exterior.

In male fish, spermiation and sperm hydration are the two important stages of maturation of male gamet. While spermiation refers to release of mature spermatozoa into the sperm duct, hydration is the dilution of spermatozoa with seminal fluid. The diluted spermatozoa is released as milt during spawning.

Environment for Breeding of Fishes

Fabricius (1950) studied the different stimuli for the spawning activities of fish and concluded these were activated by an inherited neural releasing mechanism peculiar to each species in which various external as well as internal stimuli cooperate by heterogeneous summation and the release is usually brought about by several different combination of the stimuli. Of the external factors temperature and light are considered to be the most important ones influencing maturity and spawning in fishes. Rain, flood, increase in dissolved oxygen content and pH of water are also believed to be important external factors responsible for spawning. According to Swingle (1956) fishes probably excrete into the water a hormone-like substance, or substances which may reduce or prevent reproduction. Because of this repressive factor certain fishes do not spawn when kept over-crowded, but spawning occurs when they are released in fresh water. Influence of population density on the reproduction of fish has been observed by Kawajiri (1949) and in Europe carp is bred by releasing brood fishes in the fresh water of " Dubisch ponds " (Schaeperclaus, 1933). Of the internal factors, endocrine secretions, especially that of the pituitary gland is believed to induce ovulation and spawning in fishes (Pickford and Atz, 1957). This function of pituitary is again influenced by external factors like rain, light, temperature etc.The external environment mediates its effect on the endocrine system through the pituitary (Hoar, 1957). The sexual play of brood fishes under optimum conditions also appears to stimulate secretion of these hormones.

Breeding habits differ considerably in different groups of fishes. According to their breeding habits, the important cultivated pond fishes of Asia and Far East may be grouped into three distinct groups.

Group A

This group includes fishes, such as, common carp (*Cyprinus carpio*), tilapia (*Oreochromis mossambica*), gourami (*Osphronemus goramy*), sepat siam (*Trichogaster pectoralis*), pearl spot (*Etroplus suratensis*), kissing gourami (*Helostoma temmineki*), nilem (*Osteochius hasselti*), tawes (*Puntius javanicus*), mata merah (*P. orphoides*), murrels (*Channa* spp.), which breed in ponds.

Parental care has been observed in tilapia, sepat siam, pearl spot, gourami and the murrels.Male tilapia makes a circular pit at the pond bottom at which the female lay eggs. The fertilized eggs are picked up by the female and kept in the mouth, where further development and hatching take place. The gourami builds a nest out of grass and weeds on the pond margins, the male sepat siam builds a foam nest in which eggs are laid. The pearl spot attaches eggs in layers to hard substrate. The murrels build nests with vegetation near the margin of ponds in shallow waters and the parent fish guard the fry.

The other cultivated species of this category, such as, common carp, nilem, tawes, mata merah and kissing gourami do not show parental care but breed in ordinary ponds. Nilem and tawes spawn when well-oxygenated running water is supplied to the ponds. While the eggs of nilem and tawes are demersal and non-adhesive, those of kissing gourami float on the surface and the mata merah eggs are adhesive. The common carp also breeds in ordinary ponds, but the intensity of breeding is accelerated when they are released in fresh water. The adhesive eggs are attached to aquatic plants or other objects. Majority of fish belonging to this group breed more than once or several times in a year.

Group B

Indian and Chinese carps, belonging to this category do not ordinarily spawn in ponds, but breed in flooded rivers and adjoining areas. Indian carps are known to breed in "bundhs" which are specially designed ponds receiving inflow of fresh rain waters, but Chinese carp were believed to breed only in their natural habitats in China. Kuronuma (1954) reported the breeding of silver and grass carps in the Tone river of Japan. Tang (1960) has also obtained fry and fingerlings of these carps in the Ah Kung Tian reservoir in Taiwan.

Group C

The marine and brackish water species, mainly belonging to this group, do not spawn in ponds or so far could not be induced to span in ponds. Milk fish (*Chanos chanos*), sea bass (*Lates calcarifer*), mullets (*Mugil* spp) and eel (*Anguilla japonica*) come under this group. The milk fish breeds in the sea and the eggs are pelagic. The species has not been observed so far to attain maturity or to spawn in ponds or in other confined waters. The grey mullet (*Mugil cephalus*) also does not breed in ponds or in estuaries and is believed to spawn only in the sea. The fresh water mullet (*Mugil corsula*) breeds in the fresh waters of rivers and the sea bass in brackish water rivers, creeks and lagoons. The eel spawns only in the deep sea. The young ones of most of the above mentioned species are collected from the sea coast and estuaries for stocking in ponds.

Marine Pelagic fish

All the marine animals have evolved to fit the requirements of their respective natural environments, and every species have secured its own ecological niche in order to survive. Fish like, herring and sardine that feed on planktons have a high reproductive capacity, but their ecological niche is relatively unstable and subject to unpredictable patterns of large cyclical changes. In contrast, fish higher up on the food chain like tuna and flounder are less productive but have a more secure ecological niche. Although they are not subject to large natural fluctuations, they are more susceptible to mortality caused by fishing.

Pacific herring is a cold current migrating fish that originates in the Arctic Ocean and is divided into two genera, Pacific herring, *Clupea palasii* and herring *Clupea harengus*, in the Atlantic. Pacific herring inhabit the coastal waters of both Asian and American sides of the North Pacific, with their area of distribution extending from the Pacific through the Bering Sea and into the Arctic Ocean.

With regard to location of spawning ground and migration routes, The Pacific herring is divided into a large number of distinct groups, or populations.

First of all, the populations can be divided into two major groups depending on their life cycles. These include (a) population that migrate over large offshore areas and show large fluctuations in the amount of resources, known as offshore type Pacific herring and (b) regional Pacific herring that migrate over a smaller confined area and represent only a small amount of the resources. The regional type can be further divided into lake type Pacific herring that spawn and migrate in the low salinity waters of lakes inner parts of bays or marshes and the coastal type which limit their habitat to a specific area of coastal water. The Hokkaido spring Pacific herring is an offshore type type resources that are believed to spawn along the Japan Sea coast of Hokkaido, the coasts of Okhotsk Sea and the Pacific.

Pacific herring migrate to their spawning grounds in schools in the northern hemisphere's spring months of March through May. The spawning grounds they choose are rocky and gravelly bottom areas with sea weeds at a depth of less than 15 meters. The females lay their eggs on sea weeds and the males release their sperm in the surrounding water to fertilize them. When the spawning school is large enough the male's sperm turns the water of a whole area whitish in colour.

When the larva reach a body size of 9-10 mm, they begin to move about actively and leave the sea weed where they hatched and begin migrating in search of food. Much of the life cycle after that has been proposed but is still unknown. However, with regard to the Hokkaido Sakhalin population the hypothesis of Yamaguchi, M (1928) is most widely acceped.

According to his theory, the larva that hatch along the west coast of Hokkaido migrate east along the Okhotsk coast and then out through the Kurile Islands into the Pacific where they migrate over a large area in a clockwise direction along the coasts and offshore waters from Hokkaido down to the Sanriku coast of Honshu for a period of about 2 years.

After this the fish return to the Okhotsk Sea, where one group will return to the Japan Sea to spawn the following spring as three years olds; while the other remains for one more year in the Okhotsk Sea before entering the Japan Sea the next year to join the spawning stock.

After spawning, the fish spend the summer in the offshore waters of the Japan Sea before migrating south in the fall and returning to the coastal spawning grounds once again in the spring. This pattern is then repeated every year, making the Pacific herring the object of fishery every spring.

Pacific herring are surface water migraters, by character, but in the cold water season both the immature and mature herring will move to deeper waters to spend the winter. As for feeding habits from the larva and juvenile to fry stages, herring feed primarily on the crustacean *nauplius* or the *Copepoda* and as adults they mainly feed on *Copepoda* and krill as well as small crustaceans and fish fry. Herring have a life span of about 17 years, but they only form the schools that make them viable object of fishery until about their eighth year.

Zostera Beds: The Cradle of Marine Reproductivity

Sea weed beds are the infant cradles of the sea. These beds where sea weeds or sea grasses grow densely over a sufficiently large area of the sea bottom often play a vital role as spawning and nurturing grounds or habitats for marine organisms.

Sea weed beds can be divided into a number of types including zostera beds made up mainly of sea grass as well as Sargssum beds made up mainly of gulf weed, laver beds, *Eisenis bicyclis* and kelp beds etc.

Zostera is a flowering plant that proliferates in the waters of inland seas and bays with gentle currents and sand and mud bottoms at depths upto 5-6 meters.

The principal zostera species found in Japan Seas are *Zostera marina* and *Zostera japonica Aschera*. These zostera beds with their bright green grasses give a special appearance to shallow sea areas and serve as the nutturing grounds for number of fishes, such as, sea bass, sea bream, rock fish and swimming crabs during their fry stage.

Some fish like black sea bream and black rock fish continue to make the zostera bed part of their habitat even after reaching the maturity. For such fishes zostera beds serve both as a feeding ground rich in small marine animals that live on the grass surface or species like sand worm and small shrimps, as well as a place to hide themselves from predators. Also cuttle fish and certain species of fish spawn by attaching their eggs to the surface of the zostera grasses.

Yellowtail, *Seriola quinqueradiata*

The species is found distributed widely through the tropic and sub-tropic waters. The species is found in abundance primarily in the warm coastal water regions of an area stretching from the southern part of the Kamchatka Peninsula through the aters of Japan and Korea in the southern part of the Maritime Littoral Province of Siberia.

The spawning grounds for yellowtail are found around Noto Peninsula (Lat. 37 Degree N) and the Bohso Peninsula (Lat. 35 Degree N) and waters south of these points. Particular concentration is seen in the waters of the East China Sea between 27 Degree and 30 Degree Lat. N.

Once the fry have hatched, their growth pattern involves attaching themselves to the drifting sea weeds, that gather in ocean fronts. These drifting sea weeds are fragments of coastal sea weeds that have come loose and float in the surface waters and they are a gathering point not only for yellowtail fry but for fry of many kinds of species. The yellowtail fry that gather among these sea weeds gradually drift northward with the warm currents until they reach a body length of some 10 cm. At this time they leave the current and migrate to the shore waters of central and southern Japan. Here they search out primarily bay and inland sea waters in which they feed and grow. It can be said, therefore, that drifting sea weeds function as the vehicle which distributes yellowtail resources through the Japanese waters.

During immaturity, the yellowtail remain in the coastal waters to which the sea currents brought them. But when they reach the age of 2 (body length approximately 40 cm) they leave these waters to begin migration over large open sea areas. During the spring and summer months they migrate north in search of food. In autumn and winter they migrate south in search of spawning and wintering grounds.

Upon reaching a size of about 10 mm, the young yellowtail begin to feed on animal planktons. These planktons continue to be their main food source until they reach a length of about 10 cm. But, in the mean time, after reaching a length of about 3 cm, they begin to feed on the fry of such species as sand launce, anchovy and mackerel pike. By the time they reach a length of about 8 cm, their consumption of planktons begin to decrease. When they reach a size of about 13 cm, they are feeding entirely on other fishes.

Yellowtail is a fast growing species of fish. After hatching the fry are said to double their body length every 20 days. From the fingerling stage until maturity, it has been recognized that growth rate varies from region to region depending on such environmental conditions as water temperature. But in the case of cultured yellowtail yearlings (the first year after birth) reach a length of 30-50 cm with a body weight of 1-1.5 kg, second year fish (between 1 and 1 and ½ years of birth) reach a length of 100-130 cm and a weight of 3-5 kg and third year fish (two years from birth and after) grow to a length of 150 cm and a weight of 4-6 kg.

In Japan, yellowtail culture fisheries have continued to rely on the gathering of natural fry for its seeds, without resorting to artificially induced hatching practices. The gathering grounds for yellowtail fry stretch over a long line corresponding to the warm currents on both Pacific and Japan sea sides. They generally overlap the prime fishing grounds for adult natural yellowtail.

During the period of highest production about 70 million fry were stocked each year.

Red Sea Bream

Pagrus major, the red sea bream are found the world over in coastal or continental shelf areas of the temperate and tropic zones.

The red sea bream is characterized by having their own particular native spawning grounds located in sea areas with a bottom of natural rock reefs at a depth of 50-150 meters. In the spring, when the water temperature rises to about 14 degree Celsius, mature sea bream migrate to their native spawning ground and reside there until the water temperature reaches the spawning range of 15-17 degree Celsius. When spawning commences, the parent fish rise to the surface waters in the hours around sunset to lay their eggs. After hatching, the young are carried by the currents as they float in the upper layer until they reach a coastal water area with a depth of 30 m or less. The spawning ground and nursery ground are seperated. It has long been known that red sea bream change their habitat in the different stages of growth. In infancy they live in sea weed beds in the inner waters of bays. During the juvenile stage after one year of age, they begin to broaden the range of their habitat by seeking deeper waters in the winter season when water temperature drops. After reaching 4 years of age they move out into the offshore waters, migrating widely through the middle and lower depths.

Following the life cycle from hatching through to the infancy and juvenile stages and clarifying the mechanism of resource reproduction in the natural environment is one of the most important problems involved in conducting resource propagation oriented fishery based on the release of artificially produced seeds to the natural environment.

The larva carried to the bay by sea currents, concentrate first in the eddy waters of the bay mouth area. Although some are carried further into the bay by the water currents, the majority continue their growth for some time in the middle and lower depths of the bay mouth waters.

Later, as the larva begin to take on the body shape of mature fish and their organs form, reaching the fry stage, they begin to move under their own power up into the surface water layer and beyond the barriers of the water area into the central area of the bay.

The fry then move to the lower depths of the central area of the bay and distribute themselves in spots where their primary food of this stage, copepoda, is found in abundance.

As they grow in size their main food changes from copepoda to Gammaridea and they move again, concentrating themselves in the *Zostera marina* beds of gravel bottomed parts of the inner bay area where Gammaridea proliferate.

By the time they reach a body length of about 40 mm, the young begin to extend their feeding grounds to the *Zostera marine* beds of the central area of the bay, as their feeding habits diversify to include Caprellidea and Mysedacea as well as Gammaridea.

Reaching a body length of about 100 mm, the majority of the young now extend their habitat outside the bay waters, although some will continue to spend the winter within the bay waters.

In the early stages of life of sea bream, biological and chemical characteristics of the sea environment play a large part in determining the red sea bream's habitat. As the fry and young grow, however, feeding factors play an increasingly important role, and it is recognized that the formation of large shoaling groups migrating in order to choose environments offering abundant supply of the organisms on which they feed.

In the spring season, from spring to early summer, red sea bream comes over to the specific spawning ground in comparatively shallow waters (60-199 m deep). The egg is planktonic, and larva hatches after several tens hours at water temperature of about 18 degree Celsius. Larvae and juveniles are carried over to coastal nursery ground by drifting caused by wind or the surface current. In some cases depending on the condition of ocean current, egg and larva are transported and dispersed to a fairly wide area.

It is said that the juvenile grows to the total length of 13-18 cm about 40 days after hatching and begins to inhabit the bottom layer. The habitat of the juvenile is mostly limited to the shallow water less than 30 meter deep and the bottom material of the habitat varies with the area, that is, it is sea weeds, sand or rock.

Bastard Halibut

Flat fish are a heterosomata group. Immediately after hatching they have symmetrical bodies and swim in a vertical position. As they grow, however, one of the eye gradually moves along the periphery of the head to the other side, finally becoming fixed in a position above the other eye. Its swimming posture then changes from the vertical to the horizontal, the structure of the head, bones, nerves and muscles change and it enters the benthic life with its eyeless side against sea floor and the eyed side facing up.

In the natural state bastard halibut reach an average length of 3 cm in two months after hatching and 6 cm within three months. After this growth rate increases rapidly and if conditions are favorable the young will grow to a length of 30 cm and weight of 250 gram one year after hatching, 40 cm and 700 gram by the second year and 50 cm and about 1.4 kg by the third year. At their largest, bastard halibut will reach a body length of 70-80 cm and a weight of 4-5 kg.

When bastard halibut reach a body length of about 11 mm, about 25-30 days after hatching, the left-right symmetry of their body structure begins to distort and they enter a period of metamorphosis. The metamorphosis is completed by about the 50th day after hatching. The swimming fry inhabit surface and middle layer waters at a depth of greater than 20 meter. Upon entering bottom dwelling stage at about 12 mm of body length, however, they move to sandy bottom areas at a depth of less than 20 m, most commonly inhabiting river mouth areas or areas with eddies. For the purpose of spawning or feeding, adult fish will make migrations between waters of different depths and between north and south. Prior to spawning they approach shore and live in shallow waters at a depth of 20-30 m. After spawning they will migrate in a northerly direction in search of food. When water temperatures begin to drop in the fall they will move to depths below 50 meters and begin to migrate south, where they eventually spend the winter at depths of 90 m or more. Throughout their life cycle, they consistently inhabit sea areas with sand and mud bottom composition.

Bastard halibut is a typical carnivorous fish. During the larval stage they feed on planktons, but after reaching a body length of about 3 cm, they gradually shift to a piscivorous nature. Among bastard halibuts and right-eyed flounders are varieties that feed mainly on bottom dwelling marine animals, but bastard halibut with its well developed teeth and strong swimming capability, feeds abundantly on the youngs of small and middle sized fishes that inhabit its habitat, as well as bottom dwelling crustaceans. In its natural state, bastard halibuts feeds primarily on anchovy and sand lance. In addition to this, it also feeds on the young of horse mackerel and chub mackerel together with Japanese whiting, scorpion fish and right-eyed flounders, Its feeding habits are greatly affected by water temperature. At water temperature below 10 degree Celsius bastard halibut ceases to feed altogether. Within the range of 10-20 degree Celsius it can be said that the higher the temperature the greater their feeding activity. When the temperature reaches 25 degree Celsius, their appetite begins to weaken and temperature over 27 degree Celsius they again cease to feed.

Eel

There are 19 genera of eel in the world (16 species and sub-species). Of these only two species are found distributed on the Europe and American sides of Atlantic Ocean. All of the remaining 17 are distributed around the Indian and Pacific Oceans. In the Atlantic region one eel is found only in the northern hemisphere, whereas in the Indian and Pacific Oceans they are distributed across both hemispheres, with an especially large number of species being concentrated in the tropical waters on both sides of the equator.

Eels are born in the sea, grow up in the fresh waters of rivers, lakes and ponds. The facts regarding the birth of eels, were veiled in mysterious myths from ancient times until the Medieval Ages. Although their ecology has been gradually understood since modern times, it was not until the beginning of 20th century that a Danish Oceanobiologist, Dr. Johannes Schmidt discovered some of the spawning grounds of the eel.

Eels spawn in the middle layer zone of the ocean where the temperature and salinity are high. Within 2 to 10 days after spawning, the eggs hatch. The larvae, which are transparent and leaf like, called leptocephalus.

Leptocephalus, while moving up and down during day and night, will spread out across the oceans riding on the surface of the current and will grow up as they migrate. In the case of *Anguilla japonica*, the larvae will approach the coast about one year after they are hatched (three years for European eel), transform themselves to elver (fry) near the sea bottom, and soon after transformation they will enter the rivers, and the elvers will start to ascend rivers choosing the period when the difference of the water temperature between river and coast is smallest. *A. japonica* starts the migration into rivers in late autumn to early winter when the water temperature is 8-10 degree Celsius and remain so for 7 to 8 months. In this migration, various environmental factors besides water temperature, such as, tide, sunshine and wind work to change the migration to rivers. Unlike salmon and sweet fish, eel do not have the homing instinct to their mother river. Rather it is believed that they use their sense of smell to detect fresh water and thus find a river to ascend in whatever coastal region their migration happen to bring them to.

Once the eels start living in the rivers, lakes or ponds, that are their destination, they will burrow between rocks or into holes or into the mud during the day and will go out into the waters to feed at night. The amount of feed increases from spring to autumn and eels grow up quickly during the season. The eels eat various kinds of insects, small fish, shrimp and shell fish as natural feed.

The male becomes sexually mature in 3-4 years, whereas it takes 4-5 years for the female eel to reach sexual maturity. Towards the autumn the eels which have reached maturity become bluish black on their backs, light golden on their sides, and light pinkish on their abdomens, signifying that they have reached the breeding age. They then go down the river and enter the sea at the estuary. Immediately after entering the sea, the parent eels proceed to their particular spawning grounds and they die soon after spawning and fertilization.

Sweet Fish

Plecoglossus altivelis, sweet fish is a one year fish with a life cycle that basically ends within one year. Mature fish have a body length of 25-30 cm. They are born in fall, grow to a maturity in the middle stretches of the river during the following summer and descend the river in the fall to lay their eggs in the sandy bottoms at river mouth. The eggs hatch in the river waters and float with the flow out into the sea where they spend the winter in the coastal waters. After being raised in the sea with its abundant food supply, the young begin to ascend rivers in the spring when the water becomes warmer. Almost all the parent fish die after spawning.

Inland group of sweet fish do not go down to the sea for spawning, but rather spend their entire life in lake waters.

Sweet fish grow to a body length of 25-30 cm (body weight 75-90 g) within 10-11 months and reach their sexual maturity at this time. After hatching, the fry enters sea waters where they feed primarily on copepoda type zooplanktons, at times eating larva of squids and bivalves as well. By the time they approach the coasts to prepare for river ascent, their diet become more diverse including attached algae along with zooplanktons. By the time they begin with their river ascent, they have also developed comb-like teeth on both their upper and lower jaws that enable them to pull out and eat algae, such as, *Cyanophyta* and *Bacillariophyceae* attached to the rocks on the river bottom. It is said that for normal growth, a young sweet fish with a body length of about 15 cm needs to consume 20 gram of algae (wet weight) a day.

As an eater of algae, sweet fish occupies its own unique ecological niche in the lake/river food chain different from other fish feeding on aquatic insects. The reproductive capacity of sweet fish resources in rivers and lakes depends on the algae production capacity of the bottoms of the rivers and lakes.

Sweet fish make their habitat in both rapids and pools of rivers. The middle section of rivers, where wide sections of rapids alternate with deep pools make the best habitat for sweet fish.

River sweet fish are known for the fact that individuals defend a "territory" of their own. An individual will occupy a territory of about 1 square meter in a section of rapids and feed on the algae in that area and will immediately attack any other individual that enters its territory.

Japanese fishermen have developed an unique fishing method that makes use of this trait. A live sweet fish is attached to the fishing line and allowed to swim in the rapids. When another sweet fish attack the bait sweet fish, it gets caught on a hook attached to the tail part of the bait sweet fish. The method called "decoy angling" is used by 4 to 5 million sport fishermen every year in the country's rivers.

Generally speaking, lake sweet fish can be divided into two types, one that ascend the rivers running into the lake for a period and then return to the lake for spawning, and one that spawning near the lake shores, and then, after hatching, heads out into the offshore waters of the lake where they migrate in search of food in the surface waters and grow up feeding on water flea and other zooplanktons. While the former group will grow to a body length of 25-30 cm, the later will reach sexual maturity and spawn at a body length of 5-9 cm.

Salmon

All species of this genus Salmon go through a life cycle consisting of three distinctly recognizable stages;

1. The alevin or fry stage which is spent in the river after hatching.
2. The growth stage in which the fish go downstream into the ocean and migrate for feeding.
3. The reproduction stage in which the fish return to their native rivers, seek out mates and lay their eggs.

The genus *Salmon* shows three distinct biological characteristics;

1. They are born in rivers and mature in the oceans.
2. They always return to the rivers of their birth to lay their eggs.
3. After the reproduction process all members of the parent generation die.

(a) The eggs and fry of salmon require large volumes of clean spring water. With chum salmon (*Oncorhynchus keta*), the fish that return to the river for spawning range widely from age 2 to age 6 with the largest proportion being ages 3-4. As a result, the average migrating school will consist of individuals of different ages. This fact reveals that the number of fish returning to spawn each year is fairly constant. (By contrast, returning fish in the pink salmon are always age 2. Thus environmental changes tend to cause drastic cyclic changes in the number of returning fish in alternate years).

(b) Chum salmon tend to migrate back to their rivers of birth in a fairly concentrated period of the year, and the maturation of their reproductive functions commences as soon as they begin to ascend their rivers. This means that it is comparatively easy to catch parent fish and rear them. (Pink salmon, on the other hand, begin their river ascent sporadically over a longer period from spring into summer and mature sexually while residing in the middle and upper waters of the river before laying their eggs in the fall).

(c) The eggs of chum salmon measure from 5-9.5 mm, making them second in size only to chinook salmon among the species of the salmon family. This fact makes them easy to handle during hatching operation. Furthermore, the alevin reach a body length of 2-3 cm, while feeding on the nutrients of their egg yolk in the shelter of pebbles on the river bottom. And when they emerge into the waters they already have developed mouths and digestive systems as well as considerable swimming capability. This means that they tend to have a high survival rate and are easy to feed.

(d) Chum salmon descend their rivers to the ocean within a few months of hatching. This give them access to the abundant food plankton of the sea waters and contributes to rapid growth (Cherry salmon, in contrast), habitually spend their first one or even two years after birth in the relatively unproductive waters of their native rivers. This makes them vastly less productive as a species than chum salmon or pink salmon).

Salmons which descend to the seas sometime during their life cycle, grow while migrating through four substantially different types of water environments. Fry which have descended a river first live in river mouths or inlets which are under the influence of river waters. Then they move on to the low salinity waters along the coasts. During this period, the young become acclimated to the salt water while feeding to advantage on the abundant plankton in the coastal waters. As they enter the juvenile stage in which their body shape takes on the characteristics of a mature fish, the young fish begin to venture out farther into the colder off shore waters and finally they disappear completely from the coastal environment.

In the case of natural eggs, the mortality during the fertilization and hatching stages and until the young enter the free swimming stage is very high, resulting in a survival rate of only about 10 per cent.

Artificial hatching eliminates population decrease almost completely during this period. From the time the fertilized eggs are initially incubated until the young are released into the river, a survival rate of between 80-90 per cent is said to be maintained. During this stage the following points become the object of technical concern.

(a) Securing mature parent fish and obtaining eggs that have began to mature.

(b) Egg obtaining and fertilization techniques are points of concern when moving and handling the fertilized eggs.

(c) Proper incubation of the eggs from the time of eye development until hatching (maintaining a state of rest in a darkened place, providing sufficient supply of oxygen, supplying sufficient water, preventing the growth of marine bacteria, disease prevention etc.)

The subject of concern is the survival rate after release. In 1964, a Canadian researcher calculated the survival rate during each stage of the life cycle of pink salmon. When each stage begin with a population of 100, the rate is;

(i) 7.8 per cent for the river life stage from the egg to the fry,

(ii) 5.4 per cent for the stage spent in the coastal waters,

(iii) 56.4 per cent during the stage spent in offshore ocean waters, and

(iv) 93.4 per cent for the period in which they return to the coastal waters.

The above fact characterizes that anadromous salmons have to go through extremely severe living conditions in terms of the very high mortality during their early life stages in rivers and coastal zones.

Rainbow Trout

Fish and other other marine animals do not have the ability to adjust their body temperature to compensate for changes in the outside temperature as do mammals. Therefore, it is absolutely necessary that they live in waters of suitable temperature range. Rainbow trout is a cold water species that prefers a cold water range. It is said that they are able to live within a temperature range of 0-25 degree Celsius. However, it is believed that the temperature range at which normal feeding habits and growth can be expected is 13-18 degree Celsius. And within this range the higher the temperature, the more active the feeding habits and more rapid the growth will be.

With regard to spawning and hatching of eggs on the other hand, the suitable temperature range is said to be a lower of 7-15 degree Celsius. Furthermore, throughout the life cycle and particularly during the reproductive period, it is important that the environmental conditions be as constant as possible with regard to water temperature, volume and quality.

In the 1960s there were improvements made in the area of egg gathering that led to a dramatic rise in the percentage of fertilized eggs obtained.

Unlike the fish of genus *Oncorhynchus* that spawn only once in their life time and die immediately after, the fish of genus *Salmo* spawn several times during their life span. Thus whereas salmon egg gathering is performed by cutting open the belly with a knife, egg gathering for trout uses a method in which the eggs are squeezed out of the belly portion by hand in a way that does not kill the parent fish. As a result however, the batch of eggs squeezed out of a parent fish will inevitably contain a number of eggs that have been burst by the squeezing pressure. And at the time of insemination, the spermatozoa tend to gather on the yolk protein released from these burst eggs, thus reducing the overall fertilization rate. In order to prevent this defect in the method, two techniques were proposed.

Egg washing method – Since yolk protein is soluble in certain salt solutions, washing the eggs with an isotonic solution (that is, a physiological salt solution having the same salinity as blood plasma) before insemination, results in a high fertility rate.

Air-pressure method expulsion method – In this method, a large gauge syringe is inserted in the rainbow trout's belly and pressurized air is injected to force the eggs out. This method prevents the bursting of eggs due to outside pressure, while also ensuring complete removal of all the eggs carried in the parent fish's belly. It also eliminates the fear of the fish's out skin membrane being damaged by human hands. However, because it requires careful monitoring of the parent fish's gonad maturity, inorder to be successful, it is not well suited to facilities involved in mass scale egg gathering.

As rainbow trout eggs reach the eyed-egg stage, they acquire sufficient resistance to vibration to allow them to be shipped.

Channel Catfish

The highly evolved spawning behaviour of the channel catfish (*Ictalurus punctatus*) has presented numerous problems to the fish culturists. Its relatively nervous temperament has made the fish difficult to spawn under hatchery conditions. As such, the culturist has had either to collect parent fishes from the lakes and streams just before spawning time or to acclimatize this parent fishes for two or more years.

In the selection of parent fishes problems have been experienced in distinguishing the channel catfish from the closely related blue catfish. Trouble is also faced in sexing the fish, in order to reach workable sex ratio for breeding. Learning is required to handle the fish to minimize fighting which results from pairing during breeding season and to consider the role of male for parental care of the eggs.

The preferred parent fishe of 0.9 to 4.5 kg, hatchery raised if available, with equal numbers of males and females. In pairing the fish, Clemens and Sneed (1957) concluded that males of comparable size to the female should be used, but it is required that male be slightly larger than female. Sneeds recommends that parent fishes be stocked at the rate of 333 to 444 kg per hectare of 0.9 to 1.35 parent fishes where growth is desired. Where growth is not expected, the stocking rate can be doubled. Fish should be fed 2 to3 percent of their body weight, 3 to 4 days a week when the temperature is above 12.7 degree Celsius. It is generally accepted that fresh or frozen meat or fish should be part of the diet.

Females weighing 0.5 to 1.8 kg produced about 8900 eggs per kg body weight. Channel catfish eggs are deposited in a pile which is yellow in colour and turns red as the eggs approach hatching. The fry when hatched, contains 750 to 1300 numbers per 0.14 kg.

The period of incubation ranges from five to ten days at temperature between 21 to 26 degree Celsius. Three days after hatching, the fry start to feed and move about.

Environmental Factors

Environmental factors that regulate maturation and spawning in fishes are : Photoperiod, water temperature, water quality (namely, dissolved oxygen, pH, hardness, salinity, alkalinity etc.), flooding and water current, tides and cycles of the moon, weather cycles (namely, atmospheric pressure, rainfall), spawning substrate (namely, aquatic plants, sticks, gravel, spawning mats, spawning caverns), nutrition, disease and parasites, and the presence of other fish. Environmental stimuli are received and transmitted by the brain. Stimuli of reproductive importance are routed to a portion of the brain called hypothalamus. The hypothalamus produces Gonadotropin Releasing Hormone (GnRH) and also Gonadotropin Release Inhibiting Factors (GnRIF) that is, dopamine, which is a substance that inhibits the release of gonadotropin.

Determination of Maturity in Fish

For successful induced spawning of fish, it is essential that fish must be sexually mature. The external appearance of parent fish have been used to understand the stage of sexual maturity. In some species, males change in appearance during the spawning season, for example, hooked jaw in salmon, and rough pectoral fin in cyprinids. These physical changes makes it relatively easy to identify sexually mature males. The fish where external sexual dimorphism is not clear, sampling of eggs and sperm is essential to determine the sex and maturity of the organism.

Collection of Eggs

The ovaries can be sampled with either a rigid or flexible tube (catheter). Rigid catheters are usually made from lengths of glass or hard plastic tubing. Flexible catheters are made from polyethylene or vinyl tubing. The catheters must have an outer diameter small enough to be inserted through the genital opening and sufficient inner diameter to accommodate the eggs. The leading edge of the catheter should also be smooth or rounded to prevent damage to fish gonad. To collect the egg sample the catheter is to be inserted through the genital opening and rotated with gentle pressure down the oviduct into the ovary. Forceful pressure will puncture the oviduct or ovarian wall. Sampling with a flexible catheter minimizes damage to the oviduct. If resistance is felt, the tube should be removed and reintroduced at a slightly different angle. If necessary, suction may be applied to the catheter by mouth or a syringe to draw a small number of eggs into thetube.

Eggs to Assess the Female Maturity

Seven stages of ovarian maturation were suggested for fish by the International Council of Exploration of Sea (Qasim, 1973). These have been subsequently adopted by various workers (Guraya, 1986, 1994). However, Vasal and Sundararaj (1976), Srivastava and Saxena (1996) have opted for five maturity stages while, Belsare (1962), Srivastava and Swarup, (1979), Srivastava (1980) and Lal and Pandey (1998) have assigned the maturity stage on six point scale. Even Qasim (1973) also felt for reducing the number of maturity stages to five for marine species for ease in collecting data at the landing centers.

Ovary of freshwater catfish, *Heteropneustis fossilis* undergoes marked cyclic morphological and histological changes before attaining full maturity and becoming ripe. On the basis of shape, size, colour and other histological features of the ovary, six maturity stages were recognized in the catfish.

Resting Stage (Immature)

The ovaries are small, thin, thread-like, translucent, pale or dirty white in colour with inconspicuous vascularisation. The ovaries occupy only a small part of the body cavity and ova are not visible to naked eye. Histologically, the ovary shows ovigerous lamellae having nests of oogonia and immature oocytes in the stages I and II. Gonadosomatic index was 1.22±0.46 and average oocyte diameter 102±8.6 micron.

Early Maturing Stage

Ovaries become slightly larger, thicker, opaque and are light yellowish in colour. These increase in the weight of the ovary and they occupy about half of the body cavity. Histologically, oocytes in stage III and IV are present in large numbers. Gonadosomatic index was 2.14±0.57 and average oocyte diameter was 116±6.9 micron.

Advanced Maturing Stage

There is a further increase in the weight and size of the ovaries which have a deep yellow colour and occupy two-third to three-fourth of the body cavity. Vascular supply increases and the blood capillaries become conspicuous. Immature oocytes are reduced in number while stage IV and V oocytes are present in large number. Few oocytes of stage VI are also seen. Gonadosomatic index was 4.84±0.62 and average oocyte diameter was 152±8.2 micron.

Mature or Pre-Spawning Stage

The ovaries are further enlarged occupying almost the entire body cavity. They are turgid, deep yellow in colour and a large number of spherical ova are visible to the naked eye through the thin ovarian wall. The blood supply increases considerably. Both translucent and opaque ova are present, and the ovaries attain their maximum weight. The fish becomes gravid due to ripe ova tucked inside and the abdomen becomes round. The ova are not discharged till the environmental conditions become favorable. A large number of ova at stage VII and ripe eggs are seen in the ovary. Gonadosomatic index was 7.24±0.88 and average oocyte diameter was 228±11.2 micron.

Spawning (Ripe) Stage

Ovaries are very much enlarged occupying the entire body cavity. They are turgid and yellow in colour with large number of translucent eggs. Ovarian wall is very thin, almost transparent. Eggs are present in the oviduct also and the fish spawns a number of times during this period. The ovary is now said to be in running phase. At the beginning of this phase, ova are extruded by applying a gentle pressure on the abdomen. Gonadosomatic index was 9.46±0.78 and average oocyte diameter was 248±10.8 micron.

Spent Stage

The ovaries are flaccid, shrink and sac-like, reduced in volume and have a dull colour. The vascular supply is reduced. Some unspawned large ova and a large number of small ova are present. Histoogically, the ovary shows atretic and discharged follicles along with stage I and II oocytes. Gonadosomatic index was 4.12±0.46 and the average oocyte diameter was 109±9.8 micron.

Table 2.1: Changes in Gonadosomatic Index (GSI) and Oocyte Diameter of *H. fossilis* Durinu Various Stages of Ovarian Maturation

Maturity Stage	GSI	Av. Ova Diameter (Micron)
Resting	1.22 ± 0.46	102 ± 8.6 (Wt. 62 g)
Early maturing	2.14 ± 0.57	116 ± 6.9 (Wt. 65 g)
Advanced maturing	4.84 ± 0.62	152 ± 8.2 (Wt. 78 g)
Mature	7.24 ± 0.88	228 ± 11.2 (Wt. 84 g)
Spawning	9.46 ± 0.78	248 ± 10.8 (Wt. 82 g)
Spent	4.12 ± 0.46	109 ± 9.8 (Wt. 23 g)

The egg size, shape and colour of the eggs are indicators of maturation. Approximate diameters of ripe eggs of different species of fish are given below.

Table 2.2: Approximate Diameter of Mature Eggs for different Species of Fish

Sl.No.	Species	Egg Diameter in mm
1.	Bighead carp (*Hypophthalmichthys nobilis*)	0.9- 1.2
2.	Channel catfish (*Ictalurus punctatus*)	2.3 – 2.8
3.	Common carp (*Cyprinus carpio*)	0.9 – 1.2
4.	Grass carp (*Ctenpharyngodon idella*)	0.9 – 1.2
5.	Gray mullet (*Mugil cephalus*)	0.6 – 0.8
6.	Red-tailed black shark (*Labeo bicolor*)	1.0 – 1.4
7.	Snook (*Centroponus* sp.)	0.6 – 0.7
8.	Striped bass (*Morone saxatilis*)	1.0 – 1.2
9.	Sturgeon (*Acipenser* sp.)	3.5 – 4.0
10.	White bass (*Morone chrysops*)	0.6 – 0.7
11.	Cobia (*Rachicentron canadum*)	0.6 – 0.65
12.	Asian seabass (*Lates calcarifer*)	0.45 – 0.5
13.	Scat (*Scatophagus aurgeus*)	0.4 – 0.45

Immature eggs are much smaller than ripe eggs and are usually nearly clear or opaque white or yellow, depending on fish species. Eggs that have begun to break down (reabsorb) in the ovary appear whitish in colour and under the microscope these appear irregular in shape, and the egg contents appear to have pulled away from the cell membrane. The eggs of some species (namely, striped bass, white bass, Snook and Cobia) progressively become clear or become transparent near ovulation. This process is the result of coalescence of the oil droplets into several large droplets and then into a single oil globule. Observing the position of the nucleus is a good method of determining egg development. The nucleus of an egg in the resting phase is located in the center. As the egg matures, the nucleus moves to the end (animal pole) that contains the opening (micropyle) through which sperm enters. When the nucleus is near one edge of them, the eggs are considered ripe and therefore the fish should be injected with hormone for induced spawning.

Assessment of Male Maturity

Milt can usually be stripped from males of most species when they are ready for spawning by applying gentle pressure over the abdomen between the pelvic fins and the vent. The colour of the milt is usually creamy-white to grayish white. Number of sperms in a volume of milt is extremely variable, ranging from millions to billions of sperm per ml. Creamy white milt contains more numbers of sperm per volume than grayish white milt. Sperms viability usually can be determined by observing motility with a microscope. The sperm remain active in water for a very short period of time, usually less than 1 to 5 minutes, depending on the species of fish and the temperature of the water.

Spawning Methods

Channel catfish can be spawned in ponds, pens or aquaria with or without the use of hormones. In the pond method, an equal numbers of males and females are placed in a pond at a rate of 59 to 470 fish per hectare. Forty five liter milk cans or large drums are placed in the pond as spawning receptacles so that there is one receptacle for two pairs of fish or three receptacles for four pairs of fish. For shifting fry or eggs to nursery ponds, for better stocking control, the receptacles are checked regularly and parent fishes are some times removed after spawning to minimize contamination of diseases from adult to young fish.

In the pen method, one pair of fish is placed in a small pen, one of the series of wire fencing pens built along the sides of a pond, each of which is provided with a spawning receptacle. The pen protects the spawning pair from intruding fish and gives an opportunity to select members of each pair to delay or hastening the spawning, to handle the fish needing hormone injections and to remove them immediately after spawning so that the pen may be restocked with another pair.

In the aquarium method, a pair of fish is placed in an aquarium with running water and is induced to spawn by injection of hormones. Here spawn may be obtained at a desired time and fish will not spawn naturally, some times may be induced to spawn. The aquarium method provides eggs and fry which are uniform in age and size, minimize transmission of disease from parent to offspring and eliminates predation by adults.

The three spawning methods have different applications according to the skill of fish farmer, the quality and quantity of parent fishes, and the availability of facilities. The pod method is least demanding from the standpoint of time, effort, facilities and skill. It has special use when the parent fish stock is of marginal quality or when sufficient spawning ponds areavailable to spawn large numbers of fish with limited and unskilled man-power. The pen method requires better quality of parent fshes and personnel capable of properly sexing and pairing the fish. The aquarium method should be attempted only by the skilled culturists who desires to increase the rate of production during short spawning season or when the pen and pond method is not successful.

After spawning, the eggs are left in the care of the male or removed to a mechanical hatching trough provided with running water and a system of moving paddle for agitation.

Chapter 3

Manipulated Breeding of Fin Fishes

Inducing Spawning in Fish

Two main strategies are used to induce spawning in fish. The first is to provide an environment similar to that in which spawning occurs naturally. The second strategy is to inject the fish with one or more naturally occurring reproductive hormones or their synthetic alalogs. Often the two strategies are used sequentially, the first to manipulate maturation, then the second to induce ovulation.

Environmental Manipulation

Fish have evolved in a manner to reproduce under environmental conditions that are favorable to the survival of the young. Long before spawning seasonal cues begin the process of maturation. In many fish, this can take up to a year. When the gamets are mature, an environmental stimulus may signal the arrival of optimal conditions for the fish, triggering ovulation and spawning.

Examples of environmental stimuli are changes in photoperiod, water temperature, water quality (dissolved oxygen, turbidity, hardness and alkalinity), rainfall, spawning substrate and food availability etc. A variety sensory receptors detect these cues, including the eye, pineal gland (an organ in the dorsal part of the forebrain that is sensitive to light), olfactory organs, taste buds and thermoreceptors.

Temperature

Temperature is one of the most important factors that stimulate maturation of gonads and spawning. It has also an indirect effect on pituitary by regulating the release of gonadotropin. In carps, successful spawning takes place within the range of 24-31 degree Celsius, optimum being 27.5 degree Celsius. Though spawning occurs at high temperature (35-38 degree Celsius), the percentage of fertilization and hatching of eggs can be low, as higher temperature may cause mortality of the hatchlings.

Early maturation and spawning of fish enhanced by photoperiodic regimes have been reported by several workers. However, there are also observations on delayed maturation and spawning of some species of fish in northern latitudes where photoperiod regimes are of low intensities. In India, the minor carp, *Cirrhinus reba* has been observed to attain early maturity under experimental conditions of short light periods and delayed maturation under long light periods. These instances tend to show that the requirement of light for activation of reproduction cycle varies from species to species and from place to place.

All fish, whose reproduction is cued by light, spawn at specific phases of the annually changing cycle of day length. The constant photoperiod regimes are obviously less complex than modified seasonal light cycles and thus easier to manage and use on commercial farms. Once the spawning time of a stock has been modified in this way, the fish must be maintained permanently under light proof covers and controlled light, otherwise spawning will revert to its timing under natural ambient conditions. Usually after the spawning of a stock has been advanced or delayed by 3-4 months, then the fish are maintained on a light cycle which will give phased yearly spawning. Holding 3-4 separate brood stock under different yearly light cycles ensures that eggs are available throughout the year (namely, trout). Using photoperiod control it is also possible to spawn fish more than once a year.

The use of environmental manipulation to change spawning time is a particularly attractive proposition for flatfish, bass, bream and mullet. Because very high fecundities of these fish mean that only small numbers of broodstock need to be maintained under artificial conditions.

Food

Diet has a major influence on the condition of the brood fish and the quality of gamets. Thus the most productive ponds on a farm are chosen to hold brood fish during final maturation. Additional food is given in the form of protein rich pellets, and it may allow for five ovulations each year in carp. After spawning the brood fish are kept in the hatchery for a week so that they can be fed to build up their reserves before they are returned to brood stock ponds.

Dissolved Oxygen

The hourly oxygen consumption of brood carp is around 100 mg/kg live weight in air saturated water. The minimum level of oxygen in the water must be around 6 mg/l, and it should be at saturation level at the time of ovulation (after hypophysation). A decrease in temperature will not compensate for an oxygen deficit during this phase. In consequence, the flow of water should be adjusted and mechanical aeration used in ponds, especially at night, if necessary.

Social Environment and Substrate

Cyprinid spawning is inhibited in conditions of social stress (over population, hierarchies) or in the absence of a suitable substrate for spawning. It has been shown that introduction of an artificial substrate into an aquarium will induce ovulation in gold fish. Similar principle is used in common carp spawning ponds, where the stock benefits from the addition of a grassy substrate for spawning.

Weather

Cool and rainy days are more conducive to spawning than otherdays. It has been observed that higher successful spawning is induced on cloudy and rainy days especially after a heavy shower or when the weather is cool. Accumulated freshwater and rain water give better results as they influence maturity and spawning of fishes to a great extent.

Water Flow

Constant flow or circulation of water ensures better results. Constant flow keeps the water cool on one hand and on the other it might impart a sort of stimulation to the brooders as dissolved oxygen increases in circulating water and continuous aeration enhances its supply.

Time

Spawning takes place both during day and night. For the best results, spawning at night is preferred.

Turbidity

High turbidity in stagnant water may cause mortality of eggs and hatchlings. It has been observed that fishes breed successfully in a wide range of turbidity from 100-1000 ppm.

Hormonal Manipulation

In fish, as with all higher animals, hormones play a critical role in the reproductive process. Hormones are chemical messengers released into the blood by specific tissues, such as the pituitary gland. The hormones travel through the blood stream to the other tissues, which respond in a variety of ways. The primary tissues involved in this hormonal cascade are the hypothalamus, the pituitary gland and the gonads.

Choice of Hormone

The choice of hormone depends on several things; the species, cost, availability, efficiency, medium used for preparation, hatching, larval rearing facilities etc. Brood stock holding also affects the choice of spawning technology. Marine brood stock is more likely to be kept in tanks or cages in clear, running water, resulting in constant flushing of metabolites. Hence, dopamine activity is negligible or absent in marine fish. For such fish, GnRH need not be combined with dopamine antagonists for induced breeding.

Pituitary Gland Extract

The pituitary gland is usually collected from a freshly killed or ice preserved mature fish. Among carps, common carp is the most preferred donor fish due to availability of mature fish round the year. The brain is dissected out by cutting open the dorsal side of skull and then the pituitary is picked up after the removal of the brain from the skull.

The pituitary gland produces and stores gonadotropin hormones (GtH) which play a decisive role in ovulation and spermiation. Injected pituitary material bypasses the brain-pituitary link. It acts directly on the gonad, providing the surge in blood GtH levels that normally precedes spawning.

Gonadotropins

Gonadotropins are hormones secreted by the gonadotroph cells of pituitary gland. There are two types of gonadotropins in higher vertebrates, such as, follicle stimulating hormone (FSH) and luteinising hormone (LH). Recently several studies have revealed the presence of two gonadotropins, GtH-I and GtH-II in fishes. GtH-I is connected with follicle development and vitelloggenesis, whereas GtH-II is concerned with final oocyte maturation, ovulation and spawning.

Gonadotropins used in induced spawning are 1) Luteinzing Hormone (LH), 2) Human Chronic Gonadotropin (HCG) and 3) Pregnant Mare Serum Gonadotropin (PMSG).

All gonadotropins are glycoproteins, each molecule consisting of two sub-units, such as, a and b; a subunits of all molecules are similar whereas, difference exists between b-subunits of different gonadotropin molecules.

Luteinizing Hormone (LH)

Bovine LH has molecular weight of 28500 and has 15.5 per cent carbohydrate, b-subunit has 119 amino acids.

Human Chronic Gonadotropin (HCG)

HCG is secreted by the Langerhans cells of the chorionic villi of human placenta and it appears in the blood and urine during early pregnancy after the implantation of blastocyst in the wall of uterus and reaching peak between 50th and 60th day of pregnancy. Chemically HCG molecule resembles LH having a and b subunits. It has a molecular weight of 36700 and carbohydrate content is 30-35 per cent. The b subunit has 30 additional residues at carboxyl terminal. HCG available in market under different brand names, prolan, antuitrin-5, gonadotropin, sumaah and fertizeen. Effective dose used for induced breeding is 100-2000 IU/kg of fish. Synahorin (trade name for commercial preparation) is a mixture of HCG and pituitary gland extract. HCG is measured not by weight but by biological activity in International Units (IU). It is available in sterile vials containing 2000, 5000 or 10000 IU. The unopened vial of hormone should be stored in a refrigerator at 2-7 degree Celsius. HCG is mixed with bacteriostatic water, usually supplied with the hormone. The hormone solution should either be used immediately or divied into small volumes and kept in a freezer.

Pregnant Mare Serum Gonadotropin (PMSG)

This is another gonadotropin-like molecule secreted from the endometrial cups of the uterus of pregnant mare. It is similar to that of LH, having a and b subunits, and it function as gonadotropin. Dosage is 50-200 IU/kg body weigt for female and 20-25 IU/kg body weight for male. Puberogen consists of 63 per cent FSH and 34 per cent LH.

GnRH and LHRH Factor

GnRH or LHRH was first isolated from porcine hypothalamus by Schally and Co-workers in 1971. It is a decapeptide secreted by theGnRH neurons of hypophysiotropic area of hypothalamus which regulates the secretion of gonadotropins (GtH) from the pituitary gland. The first fish GnRH to be sequenced was that of Chum salmon, *O. nerka* (Sherwood, 1993). GtH-II release is under dual control. It is stimulated byGnRH and inhibited by dopamine which is a gonadotropin release inhibiting factor (GRIF). Chang and Peter (1983) observed higher potential of GnRHa along with dopamine receptor antagonist. This finding led to the use of GnRHa with dopamine antagonist for induced ovulation and spawning. This procedure is called Linpe method. It is so named after the scientists (H.R. Lin and R.E. Peter, 1996). Various analogs of GnRH and dopamine antagonist showed different potencies. Salmon GnRH and domperidone are more effective in most of the fishes. However, in others like Atlantic salmon (*Salmo salar*) and Gill head sea bream (*Sparus aurata*) and Striped bass (*Morone saxatilis*) dopamine does not show inhibitory effect on the release of GtH-II. In such species, GnRH alone could induce spawning. Haloperidol, domperidone, metaclopramide and pimozide are commonly used as dopamine antagonist in induced breeding of fish.

Ovaprim

It is a synthetic compound manufactured by Syndel laboratories inc. Vancouver, British Columbia, Canada and marketed in India by Agri vet farm of Glaxo India Ltd, Mumbai. It consists of Salmon gonadotropin releasing hormone analogues and domperidone which is known as dopamine antagonist. Use of ovaprim has many advantages over crude pituitary extract in carp breeding, such as, a higher percentage of spawning, higher number of eggs per kg of body, higher fertilization rate, higher percentage of hatching and it is user friendly. Dose of this hormone is 0.5 ml/kg of brood fish.

Ovatide

It is a synthetic hormone manufactured by Hemmo pharma, Mumbai. It is a combination of synthetic GnRH with pimozide (dopamine antagonist). Doses of this hormone for male carp range from 0.2 to 0.25 ml/kg and for female these are doubled.

Ovopel

This is developed by Mr. Godollo in Hungary. It is a combination of mammalian GnRHa and water dopamine receptor antagonist, metaclopramide. It is prepared in pellet form, each pellet contains 18-20 microgram of mGnRHa and 8-10 mg of metaclopramide. Recommended dose is 1-2 pellet/kg of fish in rohu and mrigal.

Wova-fh

It is a synthetic gonadotropin releasing hormone analogue prepared and marketed by Wockhardt Life Science Ltd, Mumbai, India. The recommended dose is 0.5 ml/kg body weight of brooder.

Dosage of ready-to-inject spawning agents (ovaprim, ovatide, wova-fh)

Female

- ☆ Catla : 0.4-0.5 ml/kg b.w.
- ☆ Rohu : 0.3-0.4 ml/kg b.w.
- ☆ Mrigal : 0.25-0.3 ml/kg b.w.
- ☆ Fringe lipped carp : 0.3-0.4 ml/kg b.w.
- ☆ Catfishes 0.6-0.8 ml/kg b.w.
- ☆ Silver carp : 0.4-0.7 ml/kg b.w.
- ☆ Grass carp : 0.4-0.8 ml/kg b.w.
- ☆ Bighead carp : 0.4-0.5 ml/kg b.w. Mahseers : 0.6-0.7 ml/kg b.w.

Males

- ☆ All species of carps : 0.1-0.3 ml/kg b.w.
- ☆ Catfishes and Mahseers : 0.15-0.4 ml/kg b.w.

At present, both the approaches, that is, environmental and hormonal manipulation are used first to complete vitellogenesis and then to induce final maturation and spawning using hormones.

Many species of fish donor readily reproduce under certain culture conditions. Others will, but not necessarily when the farmer desires. In these cases, induction of spawning can be of great value.

Two techniques are commonly used, sometimes in conjunction with one another. The first is manipulation of the culture environment to produce some important quality in the fish's natural environment. The second is injection of hormones to stimulate spawning. The hormones may be natural hormones taken from fish or other animals, generally engineered from bacteria or synthetic analogs of naturally occurring hormones.

History of Induced Breeding

It was in 1930, B.A.Houssay of Argentina injected the extract of a fresh pituitary gland collected from a fish, *Prochilodus platensisi* into viviparous fish, *Cresterodon decemmaculatus* that resulted in premature birth of young. Following this, Brazil was the first country to develop hypophysation technique on a commercial scale. In the year 1934, Von Ihering of Brazil succeeded in inducing fishes to spawn through administration of pituitary gland extract. This was followed by Russia in 1937 where sturgeons (*Acipencer stellatus*) were induced to spawn. In India, the first attempt to induce *Cirrhinus mrigala* to spawn by the injection of mammalian pituitary extract was done by Khan, H (1037). Later Chaudhuri, H (1955) succeeded in inducing *Esomus dandricus* to spawn with pituitary gland extract of catla and *Pseudotropicus* with the pituitary extract of *Cirrhinus reba*. Chaudhuri and Alikunhi (1957) successfully induced *Labeo rohita, C. mrigala, C. reba, L. bata* and *Puntius sarana* to spawn with carp pituitary extract. In India the first marine fish induced to spawn through LHRHa was Asian Sea Bass (*Lates calcarifer*) by A.R.T. Arsu *et. al.* 1997 in Central Institute of Brackishwater Aquaculture.

Methods and Selection of Sexually Matured Fishes

The breeding methods of fishes that readily spawn in confined waters are comparatively simple. The parent fishes has to provide proper facilities for attachment of eggs, building of nests etc.Parent fishes are required to be selected considering the attainment of final stage of maturity.

Cyprinus carpio (Common Carp)

Unlike in Europe and other western countries, where the *Cyprinus carpio* breeds only in spring; in most of the Asian countries and far east countries, common carp breeds almost throughout the year and the technique of breeding and egg gathering vary in different areas.

Developed techniques of breeding and egg gathering has been evolved in Indonesia. Of several methods, the Sudanese method is considered to be important.

The method is extensively used in West Java and many other parts of Indonesia. Separate ponds are maintained for the parent fish spawning and hatching of fertilized eggs. The parent fishes are segregated sex wise and either kept in separate ponds or in some cases in the same pond isolated by bamboo screens. The parent fishes are fed with supplementary feeds like rice bran, kitchen refuse etc.

Spawning ponds are elongated and usually 25 to 30 square meter in area having a hard bottom devoid of mud and silt. The pond is dried for a few days and then filled with oxygen rich clear water. In case of turbid water, parent fish is introduced in spawning pond after sedimentation in a separate tank. The depth of water is maintained at 50 to 70 cm. Special type of egg collection mats (Kakabans) are used instead of weeds. Hair-like black fibers of *Argena* spp are thoroughly cleaned and arranges in thin layers 50 to 70 cm in length and tied together on the two sides.

In the spawning pond the egg collection mats are placed transversely on a long pole close to each other and they float at the water surface. The long pole being held in position between two posts fixed at both ends. The weight of kakabans keep them slightly under the water surface. Usually 5 to 8 kakabans are placed per kg body weight of female parent fish. The female attaches the fertilized eggs to the lower surface of the egg collection mats. Periodically these mats are inspected and turn over them when lower surface is covered with eggs.

The mats with eggs are then transferred to hatching ponds, the area of which is generally 20 times bigger than that of spawning pond. The hatching pond is also kept dry for some days and filled with water a few days before the spawning take place. The mud sticking to the kakabans is carefully washed and removed and the mats there after are placed transversely on floating bamboo poles at a distance of 5 to 8 cm from each other. Some wooden poles or banana stems are place on these poles so as to keep the egg mats about 8 cm below the water surface. The hatching pond has a number of outlets at different levels to regulate the outflow of water and gradual draining is effected when the fry are collected in sieves.

In China, submerged aquatic weeds like, *Ceratophyllum, Myriophyllum* and water hyacinth are generally used for the attachment of eggs. Square bamboo frames are placed in order to keep the weeds together in place. In general, the Indonesian technique of breeding is followed in Thailand, Malaysia and Vietnam. Kakabans are used as egg collector, but instead of the fiber of the *Arenga* spp, other fibers, such as, used for manufacturing ropes, brushes etc. are used.

In Taiwan, traditional methods are used for breeding and cultivation of common carp. In the Philippines, the common carp has been introducd in rcent years and the species is now established in several fresh water lakes and is also reared in fish ponds. It has been reported that the common carp has been successfully induced to breed by pituitary hormone injections. In Sri Lanka, *C. carpio* culture has lately developed and thousands of fingerlings are produced in the hatcheries of Fisheries Department and are mainly stocked in the major irrigation reservoirs.

In India, the original Prussian stock of common carp was introduced first in the Nilgiri Hills and was believed to breed only in cold waters at high altitudes. In recent years the strain could also be bred in the warm waters of the plains. A warm water strain was introduced from Bangkok in 1957, which is now cultivated all over in India.

The fish breeds throughout the year with the peak periods in January-March and July-August. It sexually matures in 6-8 months, the minimum size of maturity being 15 to 20 cm in length and 80 to 170 gram in weight. The same fish can be bred 4 to 5 times in a year. The male and female parent fish are segregated a few months before spawning and are regularly fed with supplementary feeds like oil cake and rice bran. No separate spawning ponds are required. The fishes can be bred in cement cisterns with tap or pond water or in hapa fixed in ponds. With one big female parent fish usually a number of smaller male parent fishes are introduced ensuring that total weight of males equals to female. The breeding hapa is made of thin cloth stiched in the shape of inverted mosquito net with tapes at the margin and fixed by means of bamboo poles in the marginal area of the pond. A thin meshed cover is put over the hapa to prevent escape of parent fishes after their introduction. The size of the breeding hapa varies from 1.8 m X 0.9 m X 0.9 m to 3.6 m X 1.8 m X 0.9 m depending on the size of parent fishes. Common submerged aquatic weeds like *Hydrilla* and *Najas* are generally used as egg collectors. About 45 kg of weeds are provided for the breeding of a female weighing 3 to 4 kg. Parent fishes are released in the evening and the hapa are examined for the eggs by the next morning and extent of spawning and percentage of fertilization is ascertained. The weeds with attached eggs are thinned out and distributed to a number of hatching hapas. Hatching hapa are also cloth containers without top cover. The usual size is 1.8 m X 0.9 m X 0.9 m. The incubation period is about 2 to 3 days. After another 2 days the fry are collected from the hapa and stocked in ponds. This technique is more or less followed all over India. In the eastern parts of India, however, breeding of *Cyprinus carpio* in ponds with water hyacinth as egg collector is also practiced.

Selection of Parent Fishes

Proper selection of parent fishes is of utmost importance for successful breeding. The fish breeders of these countries have developed their own criteria in the choice of parent fishes from the practical experience gained by breeding of carps for years. In selecting good spawners usually factors like the shape of the body, fecundity, age and health are taken into consideration. Through selective breeding quick growing strains are also produced, which are selected as parent fishes subsequently.

The criteria for parent fish selection should be weight, body shape, fin development, proper shape of lateral line, number of scales and their distribution on the body and proper ratio between width and length.

Sexes are usually distinguished by the examination of genital opening. Only the larger and older males develop some tubercles on the sides of the head and roughness on the pectoral and pelvic fins. In fully matured parent fish, however,the milt and a few eggs could be extruded by giving gentle pressure on the abdomen.

The Indonesian select female parent fish of one and half year, but preferably two years age weighing 1 to 2 kg. The males are usually one and half years of age weighing one kg. These selected parent fish are bred once in every three months for about 4 years. Sometimes a number of smaller males are introduced with a bigger female, but the ratio of male and female is always maintained as 1:1 by weight. The main criteria followed by average Indonesian fish culturists are;

1. Big fairly soft belly, the lower side broad and flattened so that the fish can stand on its belly;
2. Great body depth and high and strong caudal peduncle;
3. Relatively small head and pointed snout;
4. Rather large scales in regular rows, and;
5. Genital opening situated far back.

The central and mid-ventral position of the last scale before the genital opening is also considered as the sign of good spawner. The mode of attaching eggs to the kakabans by a female is often taken into consideration.

In Indonesia and also in Japan several coloured varieties and strains have been produced by selective breeding. Some of the Indonesian strains are, olive green, red, yellow and long-tail variety. Besides there are varieties with black, white and other colours which are not very common.

According to Lin (1950), the Chinese select female parent fishes 3 to 6 years old weighing over 2 kg with soft and large belly. Smaller active males with lesser weight are introduced along with the females for breeding.

In the Philippines, in regular carp breeding, care is taken to select individuals that are found to be fast growing and hardy.

In India, special care is taken to select healthy gravid females with soft abdomen bulging right up to the region of the vent. Such a female when kept on its belly on the ground will rest on it and the belly will sag. When the fish is held upside down a small crease will be formed between the pectoral and pelvic regions. Further, it will not have a median ridge between pelvic and anal regions, and the vent will be projected very slightly like a smooth papilla with the distal margin gently cleft at the middle. Healthy deep bodied males with relatively smaller head and soft abdomen oozing milky white milt are selected.

Breeding Methods of *Puntius javanicus*

In Indonesia, *Puntius javanicus*, a freshwater river fish, which generally spawn in rivers during rainy season, attain sexual maturity in ponds and breed there under favorable conditions.

They are bred in special spawning ponds, usually with an area varying from 200 to 400 square meter and a depth of 35 to 50 cm. The pond bottom should preferably consist of a mixture of sand and silt, though in some ponds it is covered with fine gravel. The pond is slowly filled with well-oxygenated water and the parent fishes, that have been conditioned for 3 to 5 days in running water are introduced. In each pond about 30 to 50 pairs of parent fishes that have been kept segregated are released. Spawning usually occurs at night when a humming noise is produced. In case spawning does not occur, the fish are induced to breed by beating the water with bamboo slats. The fertilized eggs settle at the bottom and are evenly spread there. Flow of water is then stopped. The eggs hatch out within 2 to 3 days and the fry are are raised in the spawning pond till they reach a length of 2 to 3 cm. This method is commonly practiced in West Java, Central and East Java and in Central Sumatra.

Another method, commonly practiced in eastern part of West Java. The spawning pond consists of two compartments. While the first compartment serves as a reservoir and sedimentation tank, the second compartment is the actual spawning pond. A large number of fish can be bred at a time by their method. A third method for breeding *P. javanicus* has been developed in East Java. The spawning pond, 0.1-0.2 ha in area is constructed in such a place that it receives sufficient quantity of rain water from the catchment area. A small pit (4 m X 4 m) is dug at the bottom of the pond into which parent fishes are released. After a heavy shower the rain water from the catchment area enters the pond and the fishes are induced to spawn. *P. javanicus* can be bred at intervals of 3 to 4 months and a parent fish can spawn a maximum of five times. The males can be used for breeding about 50 days after every spawning.

Selection of Parent Fishes

Indonesian farmers select female *P. javanicus* 14 months to 2 years old and about 300 gram in weight for breeding. The males should be at least 10 months old. Parent fishes with large bright silvery scales arranged in regular rows, high body, rounded belly, pointed snout and the genital opening placed nearer to caudal peduncle are usually selected for breeding.

Breeding Methods of *Osteochius hasselti*

In Indonesia, the species is cultivated in ponds and can be bred in special spawning ponds. The technique is highly developed in West Java. Four different methods are employed in different centers for pond breeding of *O. hasselti*.

Since the species requires clear well-oxygenated flowing water for spawning, a sedimentation tank is provided in this method in addition to the spawning and hatching ponds. The spawning pond has a gradual sloping bottom and can be drained completely. The spawning ground covered by grass is towards the shallower part of the spawning pond. Clear water from sedimentation tank flows into the spawning pond, which has an outlet provided with gauze leading to the hatching pond. The hatching pond is paved with stones covered with a thin layer of gravel.

Male and female parent fishes are segregated and conditioned for about a week in ponds having a strong flow of water, after which they are released in the spawning pond. The flow of water is increased after the introduction of parent fishes and a strong water current is maintained to induce spawning. The fertilized eggs are swept into the hatching pond by the current of water, where they spread evenly at the bottom. The eggs hatch out by the third day and the larvae are collectd within 4 to 5 days.

In Tarogong method, the spawning pond is constructed in one corner of a parent fish pond which generally has an area of 1 ha, and is completely drainable. The hatching pond (5m X 1m), the bottom of which is covered with a thin layer of gravel is situated outside the spawning pond. After the release of parent parent fishes the outlet is closed and the water level increased. Later, water is allowed to flow into the hatching pond, creating a current of water which induces the fish to spawn.

In Magek method, the spawning pond roughly, 60 X 30 m in size and maintain a depth of 30 cm of water, is fed from another smaller pond situated at a higher elevation so that water flow into it in a strong current. The hatching pond is connected with the spawning pond and also to a drain by means of an outlet. The bottom of the hatching pond is covered with a layer of *indjuk* fiber, kept in place by bamboo frames. Spawning takes place in the spawning pond and the eggs are carried to the hatching pond by the current of water.

Selection of Parent Fishes

One to two year old parent fishes are usually selected for breeding. One female parent fish can be used for breeding every six months or sometimes even after three months to the maximum of 5 to 8 spawnings. The ratio of male and female for spawning is ordinarily 1:1. Soft and large bellied females and oozung males are selected for breeding.

Breeding Methods of *Trichogaster pectoralis*

The species spawns in ponds and also in paddy fields. The male forms a foam nest in which the eggs are laid. The fertilized eggs float to the surface beneath the foam nest, and hatch out in 2 to 3 days. The larvae are guarded by the male.

Although the natural spawning season of the species is from April to October, it is bred almost throughout the year in ponds of Thailand. The fish spawns in ordinary ponds having a rich growth of submerged weeds, like, *Hydrilla, Verticillata*, several times a year. A supply of fresh oxygenated water is conducive to its spawning and the conditioning of parent fishes before introduction into the pond has been observed to be beneficial.

Spawning ponds for *T. pectoralis* in Indonesia are rather deep and provided with floating and emergent aquatic weeds, like, *Pistia stratiotes, Ipomea repens, Limnanthemum indicum* etc. Once the pond is filled with water, no additional inflow of water is maintained. The dry season is considered best for breeding as rains usually destroy the bubble nest and affect the eggs.

The species reaches maturity at the age of 4 months. Specimens more than 6 months old (100 g) are generally selected for breeding. Ratio of male to female is 1:1.

Breeding Methods of *Osphronemus goramy*

Gouramy breeds almost throughout the year and the provision of suitable aquatic plants or other material for the construction of nest is all that normally required to facilitate its breeding.

In Indonesia, special spawning ponds varying from 25 sq.m to 100 sq.m in size and 75-100 cm in depth are used as spawning ponds. The *indjuk* or *arenga* fibers, dried grass or similar materials are thrown into the pond for nest building. Sometimes conical bamboo baskets are place along the embankments about 20 cm below the water surface. Parent fishes are introduced at the rate of one set, consisting of 1 male and 1 to 3 females per 25 to 30 sq. m area of the pond. The female builds a spherical nest with a small opening at the lower end, out of *arenga* fibers and dried grass arranging them in the basket. The fertilized eggs are deposited in the nest. A fishy smell and the presence of an oily substance on the water surface are considered as signs that spawning has taken place.

Soon after the spawning, the nest is taken out and the eggs are transferred to earthen ware jars of about 10 liter capacity. The 3000 to 4000 eggs laid in a nest is distributed to 3 to 4 jars, which are placed in shed or kept floating in the pond. The water is changed at least twice a day. The eggs are hatched in 5-7 days. After the tenth day, food in the form of very fine rice bran is provided and at the age of 21 days the fry are transferred into a fry pond. A modification of this method is practised by the fish farmers of Singaparna in West Java where a small pond with a continuous flow of water is used for hatching.

In few other districts of West Java gaurami breeding and nilem rearing are carried out in a common pond (700 sq.m in size). In some cases tawes is also introduced in the same pond. Aquatic weeds like *Hydrilla*, branching stems of common bamboo and *indjuk* fiber are provided to facilitate nest building. The number of parent fishes introduced is at the rate of one female per 100-150 sq.m of water area. Parent fishes are specially fed on soft leaves of plants like, *Colocasia*, *Carica*, *Ipomea* and *Manihot* or rice bran. In about 7 to 10 days the females build the nests and spawning occurs within another 3 days.

Gourami is reported to attain maturity at the age of one and half to two years. The fish farmers prefer 3-8 years old gourami for breeding purpose. The fecundity is believed to increase with the age and varies from 500 to 4000 eggs. Selected female parent fishes have plain body colour, reddish fin rays, rounded and relatively long belly and regularly arranged scales. The males are distinguished by the presence of a hump on the snout.

Breeding Methods of *Helostoma temmineki*

In Indonesia, the kissing gourami (*H. temmineki*) is bred in specialized spawning ponds and also in freshly harvested paddy fields. The size of the spawning pond varies from 30 sq.m to as large as 700 sq.m. The spawning ponds are drained out completely and the bottom is dried up and then filled with water to a depth of 50-60 cm. Paddy straw or some floating aquatic weeds are spread on the surface of water which provide shade to the eggs and also give protection to the larvae against rain and sun. In small ponds the whole surface is covered, but in bigger ponds only marginal areas are covered.

Before introduction to the spawning pond, the parent fishes are kept in segregation ponds for 3 to 4 days. One pair of parent fish is introduced per 30 to 50 sq.m of the surface area. Spawning occurs within 18 hours after introduction of parent fish and the eggs hatch out in two days. The larvae float on the surface of the water for 3 to 4 days after which they move slowly into deeper waters. After spawning the parent fishes are transferred to the stocking ponds.

In paddy fields the breeding of kissing gourami is carried out immediately after the harvest of paddy. The bunds are raised and strengthened so as to maintain a water depth of 30-40 cm. The breeding technique is more or less the same as done in the spawning ponds. Stocking rate of parent fish is also similar to that of spawning ponds. After spawning the straw is cut and spread on the surface of water. The paddy straw gives shed and shelter to the eggs and larvae.

One to one and half years old fish are usually selected for breeding. The weight of female fish is approximately 150 gram. Each parent fish can be bred 4 to 5 times at intervals of six months.

Breeding Methods of *Oryochromis* (Tilapia) *mossambica*

The species is an eurihaline mouth breeder and breeds in freshwater, brackish water and even in the sea. The fish breeds almost throughout the year, although the fecundity is low. The fish quickly establishes itself and a high population is produced because of the protection afforded by the mother in larval and young stages.

The fish attains maturity and starts breeding when 2-3 months old. In Indonesia, tilapia mature when 9-10 cm. long, but fish of 5.5-6 cm size have been observed to carry embryo in their mouth. A female breeds even 7 to 8 times in a year. No special spawning technique is followed as the fish readily breeds in all waters. On the contrary as the fish propagates quickly, proper control of its breeding is necessary for manipulation of its population in the pond. The males make circular pits of 30-35 cm in diameter and about 6 cm deep at the bottom of the pond where the eggs are laid by the females. The pond bottom, therefore, should not be hardy or rocky. The females pick up the eggs in the mouth immediately after they are laid and fertilized.

A method practiced in Indonesia where separate spawning ponds and fry ponds are maintained. The parent fish are stocked in the spawning pond and when the eggs have hatched, the female parent fish are disturbed to spit out the larvae. The water from the spawning pond along with the fry is drained into the fry pond which is situated at a lower level. This operation is repeated every fortnight and the advanced fry are collected occasionally from the fry pond.

No special effort is made for selection of parent fish. Mature males can be distinguished by the darker colour of the body and red margins of the fins. A marked difference in the genitalia also exists. The male has two orifices, whereas the female has three. Besides the female can also be identified by the distended brood pouch in the mouth.

The main problem in tilapia culture is to find out means to check reproduction of the species for proper stock manipulation and raising fish of marketable size. Mono-sex culture where the males only are raised, gives very encouraging results, but the accidental entry of females which often takes place creates difficulties. Mono-sex culture is therefore, not very popular in the Indo-Pacific region and the culture of hybrids appears to be the solution to the problem.

Breeding Methods of *Etroplus suratensis*

The pearl spot (*Etroplus suratensis*) is primarily a brackish water fish, indigenous to India and Sri Lanka. It thrives fairly well in fresh water and is cultivated both in brackish water as well as in fresh water ponds. The fish breeds almost throughout the year in fresh and brackish water ponds. The peak period being December to February.

The species attains maturity at the end of the first year when it is 15 to 18 cm. long. No special spawning ponds are maintained. As the eggs are attached by the female to submerged objects in the same rearing pond, some hard objects like, wooden planks, slate slabs, stones, bricks and bamboo poles etc. are placed. In South India, the broad stalks of coconut leaves are also provided for the purpose. The females usually attaches eggs one by one in rows to the lower surface of the submerged object.In case of intensive cultivation of the species, the parent fishes are selected and introduced in special spawning ponds or cement cisterns provided with objects of attachment of eggs for breeding. The spawn are transferred to nurseries where they are reared to fingerling stage.

Breeding Methods of *Puntius orphoides*

Although *Puntius orphoides* has a wide distribution over Indonesia, Malaysia, Thailand and Cambodia, it is only in Indonesia, the species is cultured. The breeding technique has been developed. It can breed thrroughout the year. Construction and preparation of spawning ponds for the species are the same as those for *Puntius javanicus*, excepting that the materials for attachment of eggs, such as, *indjuk* fibers, grass, straw etc. are provided. Well oxygenated water is supplied to the spawning pond and about 8 months old parent fish weighing roughly 60 to 85 gram are introduced after conditioning for at least two days. The fish are also bred in paddy fields. For one pair of parent fish about 30 sq.m of pond area is required.

Breeding Methods of *Carassius carassius*

The Crucian carp (*Carassius carassius*) is bred in special spawning ponds by the Japanese fish farmers. Male parent fish should be more than five years old, 30 to 60 cm in length and 2 to 4 kg in weight. The female parent fish should be more than three years old, 25 to 50 cm in length and 0.8 to 3 kg in weight. For every female three males are introduced in the spawning pond. Water weeds like, *Myriophyllum* are provided for attachment of eggs.

Breeding Methods of *Channa marulius* and *Channa striatus*

Murrels, *Channa marulius* and *Channa striatus* breed in ordinary ponds usually immediately preceeding the monsoon months in India. *C. striatus* breeds in Indonesia from June to May and in Sri Lanka before and after rainy season. Parental care is practised by these fishes. The parents build nests which are circular areas made amidst marginal weeds. Fertilized eggs floats on the water surface in the nest. Both the parent guard the nest and protect the young ones till they are able to fend for themselves. No special breeding techniques are adopted for these species.

Breeding Methods of *Clarias batrachus, Heteropneustis fossilis* and *Anabas testudineus*

The group, known as "live fishes" because they possess accessory respiratory organs by means of which they are able to take oxygen directly from the atmosphere, mature in the first year and breed in ponds during the monsoon months. Usually after a heavy shower, when the adjoining area of ponds gets inundated, these live fishes migrate to those areas and breed there. The eggs of *Anabas* or the climbing perch are small, transparent and float to the surface of water, while those of the two catfishes are coloured, sticky and are attached to grasses.

Indian Major Carps

Bundh Breeding

The breeding requirements of the cultivated species of Indian carps, *Catla catla* (Catla), *Labeo rohita* (Rohu), *Labeo calbasu* (Calbasu),*Cirrhinus mrigala* (Mrigal) and Chinese carps, *Ctenopharyngodon idella* (Grass carp), *Hypophthalmichthys molitrix* (Silver carp), *Aristichthys nobilis* (Big head), *Mylopharyngodon piceus* (Black carp) etc.

are more complex than the pond breeding fishes. They do not breed in ordinary ponds. But the Indian carps have been known to breed in special type of perennial and seasonal ponds, or impoundments, locally known as "bundhs" in some parts of West Bengal, Bihar and Madhya Pradesh in India and in Chittagong district of Bangladesh.

The main characteristics of the breeding bundhs are that they receive considerable quantities of rain water with washings after a heavy shower, from the extensive catchment areas and they have large shallow marginal areas which serve as spawning grounds for the fish.

The bundhs are ordinarily of two categories, namely, perennial bundhs commonly known as "wet bundhs" and seasonal ones known as 'dry bundhs".

A typical wet bundh is a perennial pond situated in the slope of a vast catchment area of undulating terrains with proper embankments and having inlet towards the upland and an outlet on the opposite side. It has an extensive shallow area which dries up during summer, but the actual pond retains water throughout the year where an adequate stock of mature carps are maintained. After a heavy monsoon shower fresh rain water with washings from the upland area rushes into the bundh through the inlet in the form of streamlets. Thus the major portion of the bundh gets submerged and during heavy floods it may even overflow, the excess water being drained through the outlet. The shallow areas of the bundhs are the main spawning grounds of carps. These shallow areas may be situated either near the outlet or in some cases adjacent to the inlet. The outlet is protected by a bamboo fencing. The flow of water through the outlet can be controlled by plugging the bamboo fencing with straw and mud.

A dry bundh is a seasonal pond and differs from a wet bundh in usually being shallower and much smaller in size than a wet bundh. With the first shower when water accumulates in the shallow depression of the dry bundh, selected mature carps from nearby ponds are released in the bundh. Usually with the next heavy shower when the dry bundh gets flooded, fishes are stimulated and breed in the pond.

There are, however, modifications and improvements over the typical wet and dry bundhs described above and also variation exists in different regions where bundh breeding of carp is practiced.

Spawning of carps in bundhs usually occur in continuous heavy showers when large quantities of rain water rush into the bundh and the water level increases considerably. At first the smaller fishes get stimulated and migrate to the shallow areas of the bundh itself or to the inundated shallow spawning ground. Later on the bigger fishes start spawning in the same area.

Spawning behaviour of carps in bundhs is very peculiar and considerable activities and sex-play are observed at the time of spawning. During sex-play the spawners rush at a high speed, males chasing females with violent splashing of waters. Usually a female is chased by more than one male and finally a male gets at the female and mating starts. During the act, the female is held in a sort of grip by the male and the male is seen just at the surface of water bending over the female while the latter gives constant jerks with its tail and eggs can be seen coming out in a stream. The male also quivers at the time of coiling ejecting milt over the released eggs. Sometimes the pair turn upside down in the course of mating. The mating lasts for a few seconds and is repeated till the female is fully spent. After the mating is over, the parent fishes become quiet for some time and then slowly swim back to the deeper waters of the main bundh. The fertilized eggs sink to the bottom and are not visible in West Bengal bundhs where the bundh water is generally turbid, but in some of the Madhya Pradesh bundhs the eggs laid are clearly visible in the spawning area.

Collection of eggs in Midnapur bundhs are usually made 9-12 hours after the commencement of spawning. The proper time for egg collection is when the embryo shows twitching movements. Very often when spawning is continued for a long period, there is every possibility of some eggs being hatched out in the bundh itself or eggs being prematurely collected. The eggs are transported to a battery of rectangular hatching pits which are fed by drains from the bundh or any other nearby pond. Palm leaves are spread over the pits to give shade and protection to the hatchlings. The eggs hatch out 15 to 20 hours after fertilization and the spawn are sold to the pubic from the fourth day onwards. In Madhya Pradesh, the general practice is to collect the eggs from the breeding bundhs or reservoir and hatch them in hatching hapa fixed in the bundh itself or sometimes the eggs are not collected and are allowed to hatch in the bundh.

Various opinions have been advanced regarding the factors responsible for the spawning of carps in the bundh regarding the factors responsible for the spawning of carps. Khan (1945) concluded that probable factors are the rains accompanied by floods which stimulate the fishes to migrate to the spawning ground, but the fish spawns only when the temperature of the water is optimum, that is, between 24 and 30.5 degree Celsius. According to him, oxygen content of the water and other physico-chemical constituents of the flood water as such were not responsible for spawning. Hussain (1945) stressed monsoon floods as the first requisite for spawning, which was supplemented by temperature and other factors. Mookerjee (1945b) on the other hand, emphasized mainly the high dissolved oxygen content of water present in the fresh rain water is essential for the spawning; rain, flood, shallowness of the spawning ground etc, according to him are only secondary factors. He also stated that mere replacement of old water of a pond with fresh rain water induce spawning of carps.

Das and Das Gupta (1945) observed that although an increased pH value and high oxygen content in water play an important role in the spawning of carps, they have no independent significance. According them, rise of temperature to a certain dgree is essential without which "no spawning of carps is possible". Some have attributed spawning to currents that are produced in many bundhs after a heavy shower, but Das (1917) and few others have found that breeding occurred when there was no current either flowing into the bundh or out of it. It has been observed that breeding in the bundhs of Midnapur is probably stimulated by the inflow of large quantities of silt laden water from the catchment area after heavy rains resulting to an abrupt rise of water level in the bundh and changing suddenly the physico-chemical condition of water. Study of physico-chemical conditions of water of a number of bundhs showed that a highly turbid water (about 2000 ppm) with a distinct reddish colour, a low pH (6.2), a low total alkalinity (traces to less than 2 ppm) and temperature 27 to 29 degree Celsius were the conditions at the time of spawning. A primary factor of importance noted in Rangamati bundh where observations were made for about a month prior to breeding, was the abrupt increase in specific conductivity of water during breeding. Saha *et al.* (1957) after studying a few dry and wet bundh of Bankura district of West Bengal concluded that almost neutral water with low alkalinity was favorable for the spawning of carps. Dubey and Tuli (1961) made observations on the spawning of carps in several reservoirs and minor irrigation tanks in Madhya Pradesh and found that the spawning occurred only after rains in various depths of water with or without any flow, in hard as well as soft muddy soil, with a pH ranging from 7.2 to 8.2 and temperature between 26 and 33 degree Celsius. Alikunhi *et al.* (1964 b) observed spawning in both wet and dry bundhs in stagnant water in Madhya Pradesh. In dry bundhs having recent accumulation of rain water, the pH and total alkalinity were lower than those of wet bundhs in which, though there had been appreciable dilution a water was neither turbid nor so fresh. Successful spawning was observed both in cool weather after local rains and also in warm dry weather.

From the views expressed above by various observers, although it is very difficult to come to a definite conclusion as to the most important factors influencing spawning of carps in bundhs, the consensus seems to be that fresh rain water and flooded condition in a tank are the primary factors providing the stimulus to spawning and sex-play finally helps in releasing the pituitary hormones and ovulation in carps. The presence of "repressive factor" (Swingle, 1958) might also be responsible for the carps becoming refractory in confined waters. The rush of flood water in the bundh probably dilutes the repressive element sufficiently and thus spawning is precipitated.

Chinese carp which were believed to breed only in their home waters in China were first reported to have bred in rivers of Japan and in the Ah Kung Tian reservoir in Taiwan.

Inducement of Carp to Breed by Hormone Injection

Fish hypophysis, the most important of all endocrine organs is believed to secrete a large number of hormones, which regulates the various physiological activities in fish. Of these he gonad stimulating hormones (FSH and LH) exercise a decisive control over the maturation of gonads and reproduction in fishes.

The pituitary gland in teleosts is situated ventrally to the brain proper in a concavity on th floor of the cranium, known as sella turcica. It is connected to the brain by means of a stalk, called infundibulum. In many teleosts the gland is encapsulated by a membrane called duramater.

The hypophysis in fish has two distinct parts, a glandular part and a nervous part. The glandular part has three lobes, the anterior, middle and the posterior.

In fishes, the middle glandular lobe is the most important and secretes all the important hormones including the gonadotrophins or the gonad stimulating hormones. In mammals two different gonadotrophic substance have been seperated. The follicle stimulating hormone (FSH) stimulates the growth of ovarian follicles in the females and the seminiferous tubules in the male. Luteinizing hormone (LH) or interstitial cell stimulating hormone (ICSH) causes corpus luteum formation in the female and interstitial cell development in the male. All these functions have been induced in fish by fish pituitary extracts or eliminated by hypophysectomy (Hoar, 1957).

The middle glandular lobe or meso-adenohypophysis of the pituitary gland which is believed to secrete the gonadotrophins has three distinct type of cells. The cyanophills or basophills secrete the gonadotrophins and thyrotrophins. The acidophills secrete he growth hormone (Somatotrophin) and a few other hormones and the chromophobes, which are generally believed to be reserve cells without any secretory function.

In many fishes, especially, seasonal breeders, the gonadotrophin content vary according to the stage of development of the gonads. Relatively low level of gonadotrophic hormones are observed after spawning season. The potency of the gland increases as the fish gradually attains sexual maturity and is believed to reach its maximum in the fully ripe fish. The granules in the basophill cells increase in number and size as the gonads approach maturity. Garbilskii (1940) however observed that the gonadotrophic activity in seasonal spawners is low just after spawning but high for several months prior to breeding.

Pickford and Atz (1957) concluded that there is little, if any, specificity in the hormones of fish and that the pituitaries of one family or species are generally effective on unrelated species. Experiments conducted by Chaudhuri (1956) and Clemens and Johonson (1964) on gold fish demonstrated that the magnitude of response is comparatively more when the donor is phylogenetically closer to receipent. Witschi (1955) observed that the teleost hypophysis when compared to that of mammals is low in FSH and contains normal amount of LH.

There is practically no difference in the potency of glands between male and female donor fish especially when injection is given on weight basis in fish culture practices.

Hypophysis is believed to be the first link between the receptor organs and endocrine system. Hoar (1957) concludes," It seems safer to conclude that the external environment mediates its effect on the endocrine system through the pituitary" Environmental factors, such as, rain, flood, temperature, light etc influence the pituitary gland to release the gonadotrophic hormones and precipitate spawning in fishes.

The breeding success of Indian carps by administration of fish pituitary hormone has opened new possibilities in solving the problems of dearth of seed of these economic food fishes.

Brazil was the first country to develop a technique of breeding fish by injection of fish pituitary hormones in 1934 and obtained normal young ones. Since then, Brazilian fish culturists are using the technique to obtain seed of the indigeneous fishes as a part of their regular fish culture program.

Russians were next to introduce hormone treatment in fish culture. They have made extensive use of this technique for solving the problems caused by the construction of dams in many rivers in which sturgeons and other important migratory species breed. With the first success in breeding sturgeons in 1937 by Gerbidskii (1938), the pituitary treatment method is being successfully applied in fish culture for obtaining all eggs of sturgeon in the sturgeon farms situated on the lower Volga, Ural, Kura, Kuban and other rivers.

In the United States, the technique is mainly employed in the production of bait minnows and for culture of channel catfish (*Ictalurus punctatus*), a commercial as well as sport fish. Attempts have been made to induce spawning in Pacific salmon (*Oncorhynchus* spp) which are prevented from reaching their spawning grounds by the construction of dams in the rivers, but success has not been achieved so far.

In India, the first success of induced breeding of major carps by pituitary injection was achieved in 1957 (Chaudhuri and Alikunhi, 1957) and since then the technique has been used as a regular part of fish culture procedure. A similar technique has been successfully employed in the breeding of Chinese carps also.

In Taiwan, fish culture has been rejuvenated by the success obtained in spawning the Chinese carps by pituitary treatment (Tang *et al.*, 1963) and promising development towards commercial production of fish has been initiated.

Fukushima (1965) has reported successful spawning of Chinese carps by hormone injections, throughout the mainland of China and a production of 1200 million fry have been achieved in 1962.

Breeding Method

The Indian major and minor carps, such as, *Catla catla, Labeo rohita, Labeo calbasu, Labeo gonius, Labeo fimbriatus, Labeo bata, Labeo boga, Cirrhinus mrigala, Cirrhinus reba* etc., are the important pond fishes successfully bred in India by injection of pituitary hormones. The pituitary glands for injection are collected just prior to or at the beginning of the breeding season from fully matured gravid fish. Glands from immature or spent fish do not give satisfactory results. However, glands of induced bred fish, collected immediately after spawning have been observed to be potent and are effectively used. Ordinarily, glands from the same species, as the recipient fish are used, but glands from other closely related species and and also from fishes belonging to different families has given positive results. Care is taken to collect glands from fishes soon after capture, although those procured from fishes preserved in ice are also found to be potent.

Preservation of glands is made in absolute alcohol and preferably stored under refrigeration in air tight glass stoppered phials. The glands after proper preservation are weighed individually or collectively in an electrical balance. The weight is noted exactly after two minutes calculated from the time of removal of the gland from absolute alcohol. The weighed glands are again transferred to phials in absolute alcohol and stored till the time of injection, when the required quantity of glands as calculated from the weight of parent fishes to be injected are taken out, allowed to dry for a few seconds and then macerated in a tissue homogenizer with a little distilled water or 0.3 per cent common salt solution. The homogenized glands are next diluted with the same liquid, the rate of dilution being 0.2 ml per kg body weight of the parent fishes. The gland suspension is next centrifuged and the supernatant fluid is taken in a hypodermic syringe for injection. To avoid preparation of the extract every time, just prior to injection, a simplified method has been successfully tried by preserving the extract in glycerine and then storing either in an air tight phial or preferably ampuled and stored for future use.

Determination of proper dosage depends mainly on the proper stage of sexual maturity of the parent fishes. By careful selection of parent fishes and injecting an adequate quantity of pituitary gland per kg of body weight of the parent fishes 60 to 100 percent success can be obtained provided the water temperature is optimum. Depending on the condition of female parent fish, a single injection of 5 to 10 mg of homoplastic pituitary gland per kg of its body weight may give successful spawning. In usual practice, however, the female alone is given a preliminary dose of 2 to 3 mg per kg body weight and kept segregated. After six hours, a second dose of 5 to 8 mg per kg body weight is administered to the female and a first dose of 2 to 3 mg per kg of body weight is given to the males, after which the males and female are put together for breeding. Inter-muscular injections are given on the caudal peduncle or near the shoulder region using a 2 ml hypodermic syringe graduated to 0.1 ml division. BD needle no. 22 is used for fish weighing 1 to 3 kg and no. 19 is used for larger ones. For smaller fishes needle no. 24 can be used.

The major carp parent fishes are raised in ponds. A few months prior to the breeding season, 2 to 4 years old adults are collected and stocked in parent fish ponds for breeding purposes. Proper care of the parent fishes is taken and with the advent of monsoon when fishes become fully mature and the temperature of water goes down, breeding operations are started. A male parent fish can be easily distinguished by the roughness of the dorsal surface of the pectoral fin in contrast to that of a female which is smooth to touch. Fully ripe males which freely ooze milt when gently pressed on the abdomen are selected. Selection of female parent fish in prime condition is rather difficult. Ordinarily a female having soft, bulging rounded abdomen with swollen reddish vent is selected for breeding. Female parent fishes are also cathetered before selection so as to find out the condition of eggs. Healthy parent fishes weighing 1.5 to 5.0 kg are preferable, as they are easy to handle.

Selected parent fishes are collected in hand nets and weighed. The quantity of pituitary gland required is next calculated for the weight of parent fishes and the suspension prepared. The parent fishes wrapped inside the hand net is then placed on a soft cushion and injected intramuscularly. One set of parent fishes consisting usually of two males and one female is then introduced into the breeding hapa. The breeding hapa are fixed by bamboo poles in the marginal waters of the ponds.

Spawning ordinarily takes place within 3 to 6 hours after the release of the injected parent fishes. In case, no spawning occurs, a third injection with a slightly higher dose is given 10 to 12 hours after the second injection to the females. No vigorous sex-play is observed as in nature, but chasing of the female by the male with little splashing of water is usually noticed at the time of spawning. The eggs are non-adhesive, demersal and swell up considerably after fertilization to the size of a pea. Approximately 8 to 10 hours after fertilization, the embryo starts twitching movements. By that time the eggs get properly water hardened and are removd from the breeding hapa. Parent fishes are then collected and marketed after collecting the pituitary glands.

A quantitative assessment of the eggs are made from the total volume and percentage of developing eggs are estimated from samples. The eggs are then distributed uniformly to a number of hatching hapa.

The hatching hapa are tied to bamboo poles in marginal waters of ponds. About 0.075 to 0.1 million eggs are uniformly spread on the stretched bottom of the inner hapa. The eggs hatch out in 15 to 16 hours after fertilization at a temperature of 27 to 31 degree Celsius. After hatching the hatchlings escape to the outer hapa through the meshes of inner hapa. Egg cases and bad eggs are left in the inner hapa which is removed when all the eggs have hatched out. The hatchlings are left undisturbed in the outer hapa till the third day after hatching. By that time yolk gets almost absorbed and the young spawn start feeding. They are collected and stocked in nursery ponds.

The hydrobiological and climatic conditions studied during induced spawning of carps indicated that;

1. Spawning occurs both in clear as well as in turbid waters and during day or night;
2. Injected fish bred successfully when the water temperature ranges between 24 and 31 degree Celsius, the optimum being 27 degree Celsius;
3. Fish breed at a fairly wide range of pH and dissolved oxygen contents of water;
7. Cool, rainy days give better results than hot, sultry days; and
5. Rain and fresh rain water are conducive to successful spawning.

The above technique of breeding of Indian carps is more or less followed throughout the India with slight modifications suited for particular environment. In several places, where the water temperature in ponds is above the optimum the injected carps are introduced in breeding hapa fixed in flowing waters in rivers. The temperature of river water is at least 2 degree Celsius below the pond water. Attempts were made to breed carps in the air-conditioned laboratory of Directorate of Fisheries, Government of Orissa at controlled temperature (27 to 28 degree Celsius) with encouraging results.

Breeding of Chinese Carps

Silver and grass carps, introduced in India in 1959 attained sexual maturity at two years old. When they were three years old, both the species were successfully induced to breed in ponds by injection of fish pituitary hormone. Mature males of these Chinese carps are easily distinguished by the processes and ridges on the pectoral fin rays. The techniques of breeding adopted were essentially the same as followed in Indian carps. Success was achieved with homoplastic as well as heteroplastic injections of pituitary glands collected from Indian carps. A total of 9 mg per kg of body weight, injected in two doses to females gave successful results. The female usually starts oozing 6 to 8 hours after injection. While in Indian carps natural spawning occurs after injection, Chinese carps generally have to be stripped and the ova artificially fertilized. The progeny of induced bred silver carp attain maturity in the first year itself. The yearlings could be bred and young ones obtained, but in large number of cases the hatchlings did not survive. In these experiments too spawning occurred 4 to 8 hours after the second injection to the female, but the eggs laid in the breeding hapa were not fertilized and artificial fertilization had to be adopted. In grass carp yearlings, however, only the males were ripe and oozing. Two years old grass carp have been successfully bred.

In Taiwan, first success in inducing spawning of grass carp and silver carp was reported in 1963. Parent fishes were collected from a reservoir and released before injection in holding tanks with continuous circulation of water. Carp pituitary glands preserved in absolute alcohol were extracted in physiological solution and injected intramuscularly in mature spawners. Females with dialated genital papilla and flaccid abdomen and males with oozing milt were selected for breeding. Successful spawning was obtained in a few which ovulated in 6 to 18 hours after receiving single injection of one whole gland collected from a carp of comparable size or after receiving a second or third injection of the same dose. Many did not respond even after repeated injections.The males received a lesser dose of 0.5 to 1.0 pituitary gland. Artificial fertilization of eggs with sperm by hand stripping was followed in all successful cases with a single exception of one bighead which was found to ovulate spontaneously in a pond 18 hours after receiving a single injection. The dry method of artificial fertilization was adopted.

About 33 percent success was achieved. The hatching rate in a solitary case was as high as 95 per cent, but in the majority of cases only 20 per cent or less hatching was observed. The reason for this was mainly attributed to over dose of hormone injected. Lowering of the temperature was believed to be one of the most important environmental factors conducive to induce spawning.

Within a year of the initial success improvements were effected in the breeding and hatching techniques employed for inducing spawning of the Chinese carps and commercial production of their fry was started in large number of hatcheries since 1965.

Experiments with fish pituitary combined with chorionic gonadotrophin gave more effective results than administering fish pituitary alone. With this improved method the percentage of success was raised from 33 to 78, thereby achieving an increase of 45 percent success. A number of brands of ganadotrophin tried of which Puberogin (contains 250 IU/CC of CG), Gonagen and Synahorin (both contain CG mixed with a little mammalian pituitary extract) gave encouraging results.

A circular basin with a steady strem of circulating water, which eventually flows out through an opening in the center was found to be very efficient for hatching.

Lin (1964) succeeded in spawning mud carp *Cirrhina molitorella* and the technique followed was almost identical to that of other Chinese carps. He succeeded by injecting glands from the skipjack tuna. Positive results were also obtained by injecting 500 rabbit units of cattle anterior lobe extract combined with fresh toad pituitaries. In all cases the breeding of Chinese carps, artificial fertilization of hand stripping was adopted.

Lin (1964) has succeeded in breeding silver carp as early as two months before the normal breeding season. This prolongs the growing season of fish and is a definite advantage to the fish farmers.

Breeding of Catfish

Majority of cultivated species of catfish breed in ponds. Induced breeding by pituitary treatment could be suitably tried in a few species which do not breed in ponds or for hastening pond breeding.

The large sized catfish *Wallago attu* which is common in ponds and in larger impoundments in India and Pakistan breeds in rivers. This fish has been reported to spawn in bundhs, but so far no attempt has been made to breed the species by hormone injection. The smaller species, *Ompok pabda* which is primarily riverine fish and is cultivated in ponds has been induced to breed in cisterns by the injection of 0.4 to 0.5 of a pituitary gland collected from the carps rohu and mrigal.

The Siamese catfishes, *Pangasius larnaudi* and *Pangasius sutchi*, also cultured in Vietnam, do not breed in ponds and their fry are collected from rivers. Large-scale breeding of these fishes by hormone treatment has not yet succeeded although Tubb (1958) has reported early spawning of *P. sutchi* accomplished by hormone injections. Two other cultivated species, *Clarias macrocephalus* and *Clarias batrachus* readily spawn in ponds and especially the later species can be bred by simple replacement of old water by fresh rain water. As such there is no necessity for hormone treatment in these species for breeding. But in certain cases the pituitary extract is used for getting early spawning of these species. Tongsanga *et al.* (1963) achieved early spawning in mature catfish *C. macrocephalus* by giving injections of 13 mg, 26 mg and 39 mg of homoplastic pituitary glands per fish 90 to 190 g in weight. The higher doses of 26 mg and 39 mg gave cent percent success.

Maturity and Breeding of *Ompok bimaculatus*

Ompok bimaculatus matures at the end of the first year. The maturity cycle of the species through various months of a year indicates that, during November-January gonad attains stages I and II of maturity. Most of them attain stage III of maturity by March. Fully ripe female attains stage III during late May and end of July. The sex of fish can be determined from the external features, by the serration of the pectoral spine which is pronounced in male and very feeble or absent in female. The fecundity of the fish varies from 10000-20000 ova/100 g fresh weight of female, and the number of ova per gram body weight of the fish varies between 100-200 numbers.

The ovarian eggs of *O. bimaculatus* are of uniform size, brown coloured and was between 0.858-1.365 mm in diameter.

The fully ripe females and males of *O, bimaculatus* were selected and induced bred using Ovaprim (GnRH based hormone) at the rate of 1-1.5 ml/kg of female and 1.0 ml/kg of male. Following injection of both sexes, breeding was achieved by stripping method as well as hapa breeding in the same way as done for carps, after six hours of hormone injection. In the case of stripping of males, they were sacrificed and sperm suspension was obtained by finely chopping the testis and macerating it thoroughly. The sperm suspension was sprinkled over the eggs for fertilization. After proper washing, fertilized eggs were kept in fiber glass tank with proper aeration facility as well as in hapa fitted in a pond for embryonic development and hatching. The embryo hatched out and the period of incubation at water temperature of 25.5 to 29 degree Celsius was between twenty two to twenty four hours.

The diameter of fertilized egg was 1.712-1.021 mm with reddish brown colouration. The larvae was also brown with 4.268 mm in size. The post-larvae was 6.039 mm, less translucent, brownish, melanophores minute and numerous and almost uniformly distributed.

Breeding of Trout

Natural breeding of trout takes place in wild waters. Prior to spawning, trout strive hard to improvise their breeding grounds, called "redd". They show chasing behaviour. Male chases the female and ultimately the female releases the eggs in the redd after complete exhaustion. These are later on ferti;ized by release of milt by male. The negative aspect of natural breeding is less percentage of survival, the reasons for this being predation, heavy flushing of water, turbulence, silt and ineffective fertilization.

The breeding season of rainbow trout starts from November and lasts up to the month of March. Snow or rainfall helps in the maturation of gonads but these minimize the spawning time. The giving of a well balanced diet to the brooders prior to spawning is a prerequisite. The fishes show prominent morphological signs during breeding season. The lower jaw of the male bends more inwards than before, while females show bright colouration. Even in confined embankments (raceways) fishes show chasing behaviour as in nature.

Segregation

The first step, before taking up breeding of trout, ripe fishes segregation. The fishes are lifted with the help of a net and examined for ripeness by pressing the abdomen. The segregated fishes are stocked in separate tanks after confirmation of status of milt from male and eggs from female. Extreme care is taken to prevent any handling damage. During this time, the fishes after collection are dipped into 2 per cent potassium permangate solution. This prevents infection. For better holding the fishes from the caudal end, mesh cloth is used as a precaution. The fishes are starved almost 7-10 days before stripping. This prevents the formation of faeces, which may otherwise can contaminate eggs.

Collection of Eggs

The required number of female fishes are taken out of the brood tank and kept in a separate tank in the water channel for easy access. The fishes are dipped in benzocaine or MS 222 (2 per cent v/v), but the experience showed that stripping without anesthetics is very successful. This is because the conscious fish oozes out eggs by jerks and the remnants of eggs in conscious fish are very less or nil as compared to unconscious one., leading to mortality in post-stripping stage. The eggs are collected in a basin by pressing the abdomen of the fish from post lateral position to the vent. The fishes after egg collection are dipped into 2 per cent potassium permanganate solution and released back in raceways and termed as "spent". After stripping 2 or 4 females, the same procedure is followed for males. Milt is collected and mixed with eggs. A ratio of 2 females : 1 male is followed as a thumb rule. The fecundity of a ripe female is 1000-2000 eggs per kg body weight. The first maturity is attained at 2 years of age. The best age for stripping is 3 years.

Water Hardening

After collection of eggs and milt the content of the basin are mixed with cock's feather or by gentle stirring of the basin, so as to achieve maximum fertilization. While collecting the genital products, there are chances of blood aliquots or faeces getting into the basin. These are removed by manual picking. Thereafter the deemed fertilized eggs are put in bucket along with water. The stripping is done by dry method, but the water which is used along with eggs serve two purposes. It hardens the outer shell of the eggs and prevents them from any physical or chemical stress. The water also neutralizes the effect of toxins, chemicals or faecal contamination, which might otherwise spoil the eggs.

Loading of Trays

The deemed-to-be fertilized eggs are kept in a bucket for 3-4 hours prior to their loading in trays. The fertilization of eggs takes almost 24-72 hours and after fertilization the egg is called "green ovum". The green ovum has a diameter of 4-5 mm. The eggs are taken out of the bucket with the help of a standard container which contain 8000 to 10000 eggs. The eggs are then evenly spread in perforated hatching trays measuring 0.51 x 0.33 x 0.10 m. The trays are arranged in sets of 4 or 7 in two sizes of hatching troughs. The dimension of a small trough is 2.61 x 0.35 x 0.17 m and that of a longer one is 4.09 x 0.35 x 0.17 m. The water is continuously allowed to flow through these troughs with a velocity of 3.0-5.0 liter per minute. This rate of flow provides sufficient oxygen to the eggs for hatching.

Eyed Ova Stage

The green ova stage is followed by "eyed ova stage". The eyed ova are little darker in colour with a dark black spot on it.The yellow colour turns to light brown after reaching eyed ova stage. Usually it takes 21 days to transform from green ova to eyed ova. The rearing of green ova in different temperatures showed variation in hatching time. Eggs reared at a temperature of 10-13 degree Celsius hatched within a period of 15 days, whereas the eggs reared at 6-8 degree Celsius hatched in 25 days. Lesser the temperature, longer is the period of hatching. The eyed ova are 3-4 mm in diameter. The water flow for eyed ova stage is maintained at 3.0 to 5.0 liter per minute.

Alevin Stage

Eyed ova stage is followed by "alevin" stage. This stage is also called yolk sac stage. Alevin measures 1.3 to 1.6 cm. The time needed from eyed ova stage to alevin stage is almost 15-20 days depending on the temperature. At 12 degree Celsius, the hatching period is reduced to 10 days. In alevin stage, the eyed ova modifies into a miniature fish with a globule of fat (yolk sac) attached at the abdominal portion. The yolk sac provides a natural food for the growth of fish. It lasts for 7 to 15 days depending on the temperature. A moderate temperature of 12 degree Celsius speeded up the yolk absorption and complete absorption take place within 10 days. Care need to be taken to avoid accumulation of dead alevins at this stage. Continuous monitoring and washing of troughs, removal of dead alevins and treatment with 1 per cent potassium permanganate solution at an interval of twice a week is necessary. The sieving of yolk sac through 2-3 mm mesh helps to filter alevins from dead eggs and fishes.

Fry Stage

Alevin stage is followed by fry stage, which measures 2.0-2.5 cm. The alevins, after complete absorption of yolk, transform into first feeding fry stage, which is morphologically alike to the adult trout, except that it lacks proper energy to move. It takes almost 10-15 days for the alevins to transform into the first fry stage. The twitching movement starts immediately after complete absorption of the yolk. The fry tends to stay at the bottom with meager movements. At this stage of their life, the fish are supplied with starter diet. The artificial diet given to the fry are also supplied with cod liver oil for providing sufficient energy to the fry for jerky movements and for movements against the current.

Breeding of *Piaractus brachypomus*

Successful breeding and seed production of *Piaractus brachypomus in captivity was done by hypophysation* in West Bengal by Department of Fisheries. The fish has a narrow plate shaped dark body, blackish in the upper part with blunt teeth. For breeding purpose parent fishes were collected locally and raised in tank along with common carp. Regular feeding was done at the rate of 6 per cent of the body weight at the evening after sunset. Boiled peas with fish meal and ground nut oil cake was used as supplementary feed. Aeration was provided by means of pump. Fishes were matured and ready for breeding after 3-4 years. Weight range of 3.5 kg for male and 5 kg for female is preferred for breeding.

During breeding season sex can be easily differentiated by some external features. Swollen belly with reddish round vent of the female and a pointed vent with depressed belly can be selected for breeding. Average fecundity of the female is between 3500-40000 eggs per kg body weight.

Breeding Method

Parent fishes were collected one day prior to breeding and kept separately sex wise in hapa. No feed was provided during this period. 1:1 sex ratio was maintained for breeding. Pituitary gland extract was used for hypophysation. The female was administered an initial dose of 2 mg per kg of body weight; while the male was not injected for initial dose. After 6 hours of the first injection to the female, the final dose of 12 mg per kg of body weight was injected to the female and 2 mg per kg of body weight was given to the male. After 6-7 hours of injecting the final dose, both the parent fishes were stripped and fertilization was done using dry method. The fertilized egg was transferred to the circular hatchery for hatching. After two days hatchlings were removed from hatchery to the nursery.

Breeding of Silver Pompano (*Trachinotus blochii*)

At Mandapam Regional Center of CMFRI, successful broodstock development, induction of spawning and fingerling production of silver pompano was achieved for the first time in India. Pompano of 250 to500 gram weight, collected from wild were stocked in sea cages of 6 m diameter and 3.5 m depth. The fishes were fed *ad libitum* once in a day with trash fish. In April 2011, four numbers cage reared adult pompano (1 female and3 males) were selected and transferred to an indoor FRP tank of 10 cubic meter capacity with photo period control facility (14 L: 10 D) for preconditioning the fishes to induced spawning. The size of the female was 39.8 cm in total length weighing 2.245 kg. The total length of the males ranged from 30.7 to 35.7 cm and weight from1.750 to 2.1 kg. The parent fishes were fed *ad libitum* with squid meat and fish roe once a day. Water quality was maintained by providing a flow-through system throughout the period. Periodic cannulations were carried out to assess the maturity of the fishes for induction of spawning. On 5th July 2011, during the cannulation of the female, intra-ovarian eggs of diameter above 500 micron were observed. The maturity of the males was assessed based on milt quality. On the same day at 14:30 hours, the parent fishes were administered with HCG at a dose of 350 IU per kg body weight. Spawning was recorded at around 4:40 hours on 7th July, 2011, approximately 38 hours after the injection. The total number of eggs spawned was estimated to be 1.30 lakh. About 50 per cent fertilization of these was

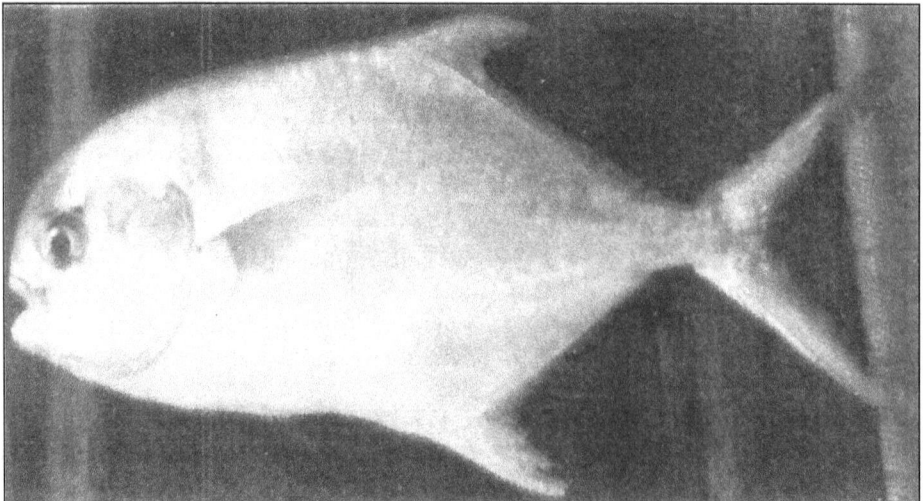

Figure 3.1: A Pompano Brooder.

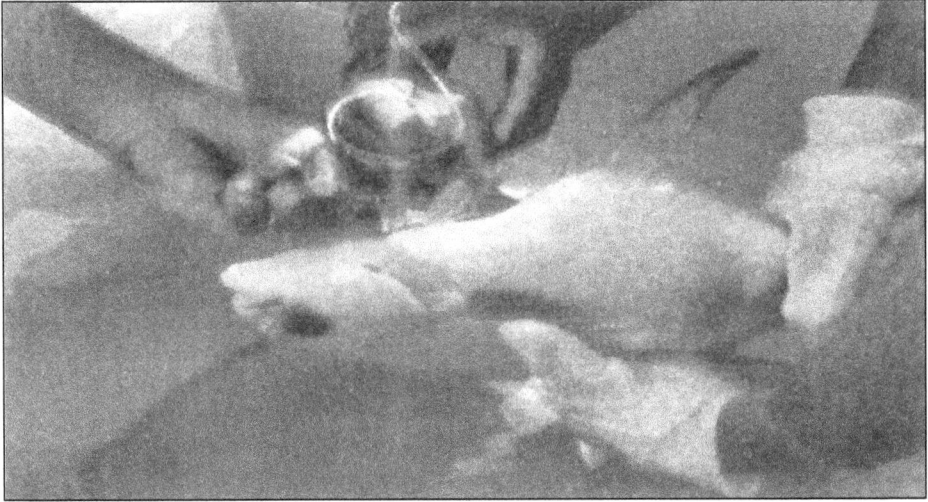

Figure 3.2: Cannulation.

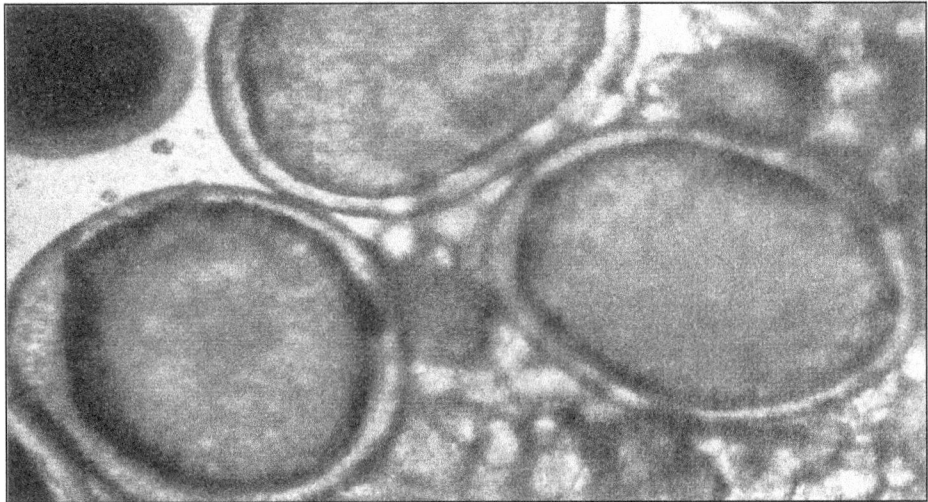

Figure 3.3: Intra-ovarian Eggs.

noted (fertilized eggs were estimated at around60000 nos.) The eggs were collected by 500 micron mesh cloth and stocked in incubation tanks of 2 tonne capacity. The diameter of just fertilized eggs ranged from 0.85 to 1.0 mm. The eggs hatched out after 18 hours of incubation at a temperature range of 30-31 degree Celsius. The newly hatched larva measured 2.0 mm in total length.

The newly hatched larvae were stocked at a density of 10000 of them in FRP tanks of 2 cubic meter capacity filled with 1.5 cubic meter filtered sea water. The tanks were provided with mild aeration and green water at a cell density of 100000 per ml. Sufficient copepods were also introduced into the larviculture tanks to provide enough nauplii to facilitate the first feeding of the larvae. On 3 dph (day

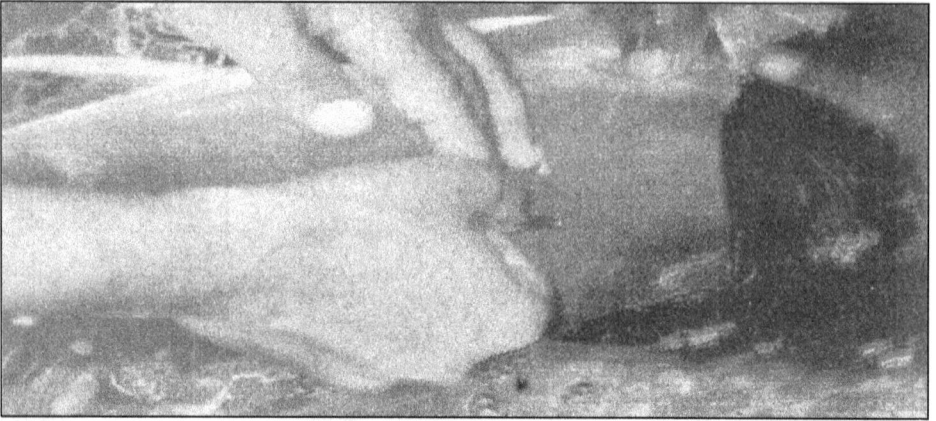

Figure 3.4: Hormonal Administration.

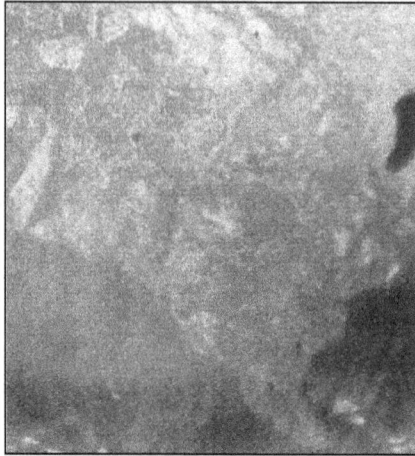

Figure 3.5: Eggs Sieved from the Spawning Tank.

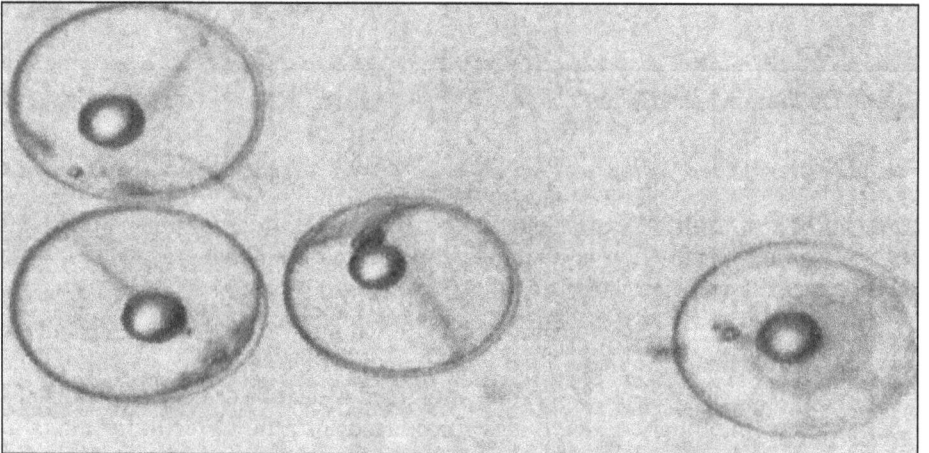

Figure 3.6: Fertilized Eggs

post hatch), mouth opening was formed which measured around 230 micron. The larvae were fed from 3 dph to 9 dph with enriched rotifers which were provided at a density of 5-6 nos. per ml in the larviculture tanks.Co-feeding with enriched *Artemia* nauplii was done during 10-13 dph and thereafter up to 19 dph with enriched *Artemia* nauplii alone by maintaining a density of 1-2 nos. per ml in the larviculture tanks. Weaning to larval inert feeds was started from 20 dph and the same was completed by 24 dph. From 25 dph feeding entirely on larval inert feeds was started. The metamorphosis of the larvae started from 18 dph and all the larvae metamorphosed into juveniles by 25 dph. During 20-25 dph gradings were done to separate the shooters. It was also noted that after the critical stage mortality during 3-5 dph, the subsequent mortalities were rather negligible.

Breeding of Cobia (*Rachycentron canadum*)

Cobia (*Ranchycentron canadum*) also known as lemonfish or ling, occurs in tropical warm waters worldwide. It reaches an average total length of 1.3 m, but some are known to reach 2 m total length and 68 kg in weight.Cobia is one of the most highly favoured food fish for human consumption. It is also a recreational species in many countries, including U.S. The artificial breeding of cobia was first recorded in Taiwan in 1992. The technique for its mass seed production was developed in 1997.

Characteristics of Cobia

Cobia is a migratory pelagic species. The range of temperature of the water in which it is found is 16.8 to 32 degree Celsius. The salinity range for cobia is found to be 22.5 to 44.5 ppt, but adoption to lower salinities seems possible (Shaffer *et al.*, 1989).

Eggs and larvae are usually found offshore, but early juveniles are more common in shallow, coastal waters, near beaches, islands, bays and river mouths. Adults are most commonly found in coastal and continental shelf waters, sometimes in estuaries. Adults are pelagic, but they often stay close to structures (buoys, drifting objects, pilings).

Although there is only one species of Cobia, there is evidence that there are distinct groups or sub-populations. Examples are the Chesapeake Bay population and the Gulf of Mexico population. This was concluded from the recaptured tagged fish (Richards, 1977 in Shaffer *et al.*, 1989).

Life History of Cobia

Cobia is gonochoristic. External differences between sexes are limited. When mature, cobia females have a higher condition factor than the males, having a full soft belly compared to the more slender, hard body of the males.

Cobia is a multiple batch spawner and the spawning season is extended over several months. Depending on the location, spawning takes place between May and August, Spring and Summer (Chesapeake Bay; Gulf of Mexico) but possibly also year round (Indian Ocean).

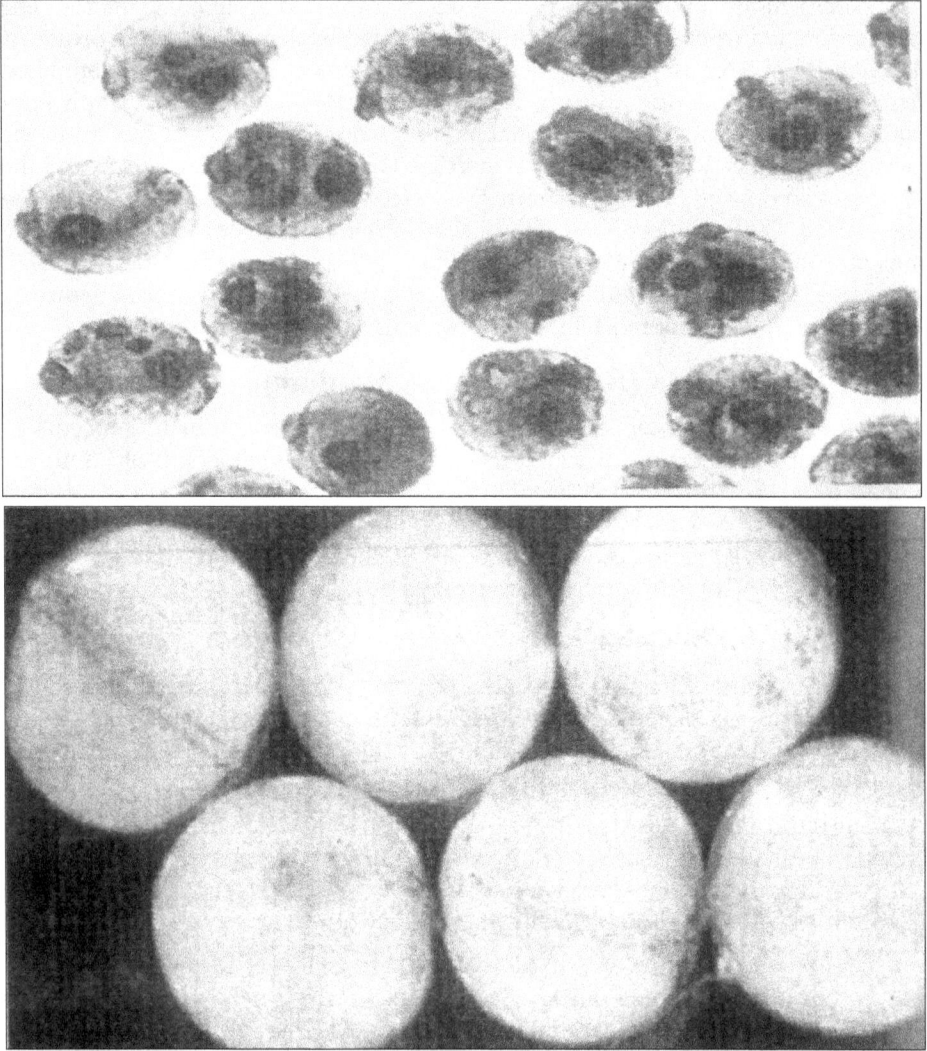

Figure 3.7: Two Microscopic Views of Cobia Eggs.

During the spawning season, the adults migrate away from the shore and form spawning aggregations. Eggs and sperms are simultaneously released and fertilization is external. Spawning Cobia and fertilized eggs have been found 40 to 80 km off the coasts in the Gulf of Mexico as well as in North Carolina waters.

The fertilized eggs are pelagic, with a large oil globule, yellow in colour and mottled with melanin pigment. They take about 1.5 days to hatch.

The larvae are 3 to 3.5 mm long and colourless immediately after hatching, carrying a yolk sac. Five days after hatching, the larvae absorb the yolk sac and reach a length of 4-5 mm and start exogenous feeding. They have limited swimming abilities at this stage.

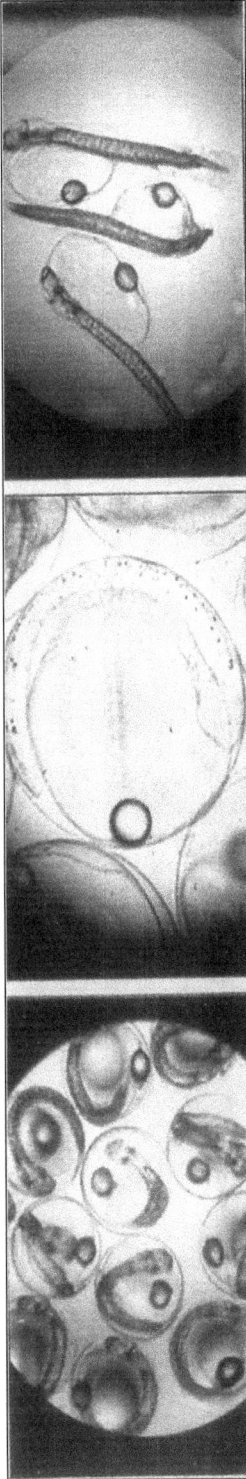

Figure 3.8: Fertilized Cobia Egg have an Average Diameter of 1.3mm and an Incubation Period of Approximately 30 Hours at 26ºC. Newly hatched larvae measure in excess of 3 mm.

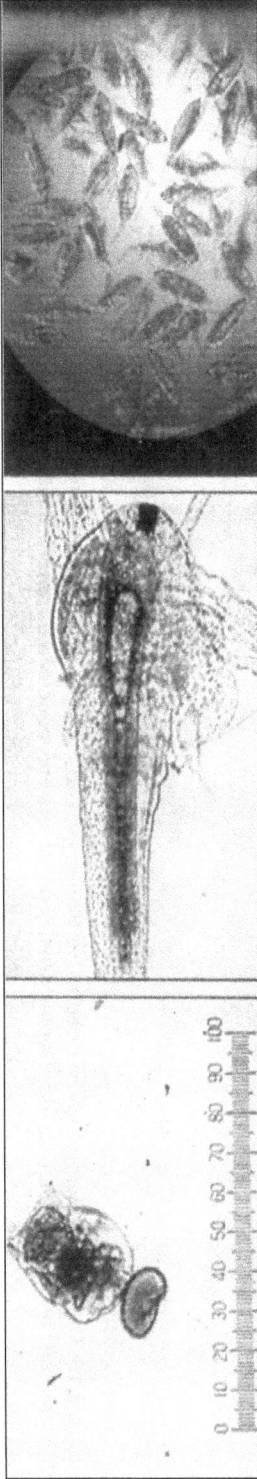

Figure 3.9: After Three Days the Yolk Sac is Used up and it is Vital to Supply Exogenous Feeds. Over the next few weeks the larvae has to be weaned on to successively larger sizes of zooplankton and finally on to dry feed. These zooplankters are rotifers, artemia and copepods (Pictured left to right above).

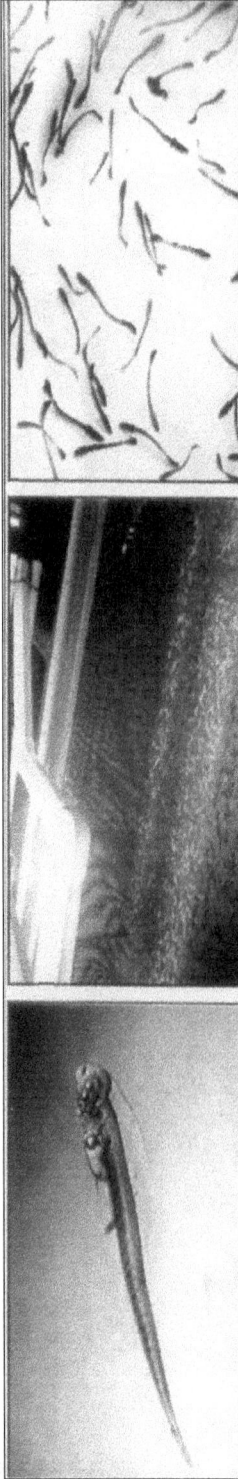

Figure 3.10: The Fast Growing Larvae need to be Monitored Closely; Water Temperature in the Larval Rearing Tanks is around 26ºC. With ambient temperatures this high, water quality problems can accelerate rapidly, if left unchecked.

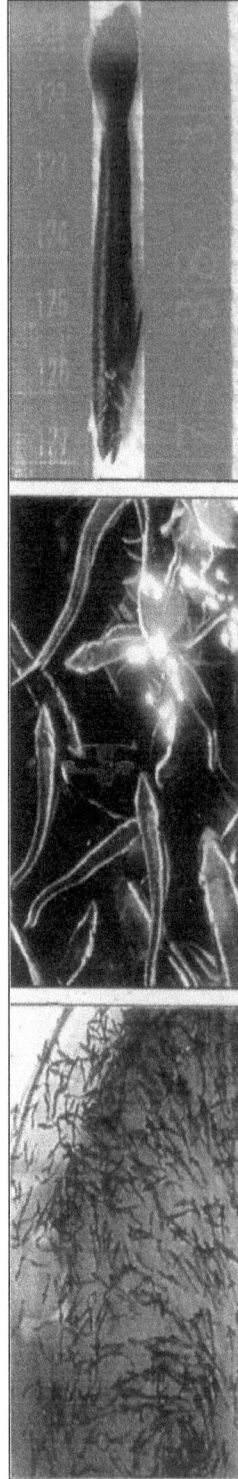

Figure 3.11: The Juveniles would have Reached 1g in Five to Six Weeks by which Time they are Ready to be Shifted to Nursery Site.

During the first 30 days, the larvae undergo metamorphosis and they start looking like the adults at a later age. After metamorphosis they look brown in colour with a black and white stripe running from snout to tail. Compared to adults, the head is more depressed and the caudal fin is large and fan shaped instead of lunate.

After about 2 months of larval life too, the shape of the juvenile remains the same but the coloured bands along the body become more pronounced; one of these is a black central line along the lateral surface, bordered by 2 gold or white bands.

When the fishes reach adulthood, they live solitarily, or groups of 2-8 fish. Only during the spawning season do the fish congregate in larger schools.

Egg Development and Fecundity

Richards (1967) and Shaffer *et al.* (1989) described three stages of egg development; (i) Immature : clear nucleated cells, 0.1-0.3 mm in diameter, (ii) Maturing : eggs with a clouded appearance, the oil globule vaguely discernible, 0.36-0.6 mm diameter and (iii) Mature : clear transparent eggs of 1.09-1.31 mm diameter with a 0.29-0.44 mm diameter oil globule. When the eggs are fertilized, they slightly increase in size. Fish that spawn in the start and end of the spawning season, were seen to produce smaller eggs than the ones that are spawned in the peak of the season (Su *et al.*, 2000). Non-ovulated eggs range in size from 0.5-0.9 mm with an egg count in the order of 2300-2800 eggs per gram (0.36-0.43 mg per egg). 18 months old females cobia had well developed gonads and a GSI (Gonado Somatic Index) in the order of 12.5 per cent. A 15 month old female cobia 8.5 kg carried 1.4 million eggs (160000 eggs oer kg body weight). Under the same condition one year old male of 7 kg was found to be mature but with a relatively low GSI (Su *et.al*, 20000.

Spawning

Cobia is a multiple spawner, releasing several batches of eggs per season. In Southern Taiwan, the spawning season is longer, running from February to October, with the peak in May. Under hatchery conditions in Taiwan, cobia started spawning in the end of February, when the water temperature reached 24 degree Celsius. Natural spawning was noticed under hatchery conditions in Taiwan. Two days before spawning the food uptake was reduced. On the day of spawning food uptake returned back to 50 per cent of the normal uptake. At 14.00 hours a ripe female was chased by several males, seemingly part of the courting behaviour. The females showed a tendency to seek shelter (Su *et al.*, 2000).

Earl (1880), Shaffer *et al.*, 1989 were the first to succeed in the artificial fertilization of Cobia eggs. Fertilized eggs range in size from 1.16-1.42 mm with an oil globule of 0.34-0.44 mm. Su *et al.* (2000) found that fertilized eggs are 1.3-1.4 mm in diameter during peak of the season and 1.3-1.35 mm early and the rest of the season. The fertilized eggs are spherical, buoyant and transparent with a yellowish tinge. The number of eggs are estimated to be 500000 numbers per kg, 2 mg per egg. Hatching was found to take 36 hours. The optimum temperature at hatching was found to be 26.5 degree Celsius and salinity was found to be 33-35 ppt. Hatching of fertilized eggs took place in 30 hours at 24-26 degree Celsius and 24 hours at 28 degree Celsius.

Spawning efforts in the USA have also been successful, utilizing round fiberglass tanks 5.5-6.0 m in diameter and 1.5-1.8 m deep to hold adult cobia. The tanks have egg collectors and are either operated as recirculating systems, flow-through, or a combination of both, depending on the biological filtration capacity of the system. Broodstock collection generally involves capturing and transporting juvenile or adult wild-caught cobia (often during their natural spawning season) into the tank systems, where 2-3 year old fish will spawn either naturally or after being induced with photoperiod and temperature manipulations. Research on maintaining and extending the cobia spawning season in the USA has resulted in the production of fertilized eggs during 10 months of the year thus far, with the goal of realizing year-round egg production in the future.

Cobia seed stock used for large scale commercial aquaculture production comes exclusively from hatcheries.

In India, at Mandapam Regional Center of CMFRI, broodstock development of cobia in sea cages was achieved by feeding with broodstock diets. The sexes were determined by cannulation and males and females were segregated and stocked in separate cages. The cannulations of female were done at regular intervals to assess the size of the intra-ovarian eggs. On 11th March, 2010, one of the females with intra-ovarian eggs around 700 micron was selected for induced breeding. The size of the female was 120 cm in total length and 23 kg in weight. Two males were also selected from the male cage. The sizes of the males were 100 cm and 103 cm in total length and weighed 11 kg and 13.5 kg respectively. The selected brooders were introduced in a 100 tonne roofed cement tank with about 60 tonnes of sea water on the same day. At around 1300 hours, the brooders were induced for spawning with HCG at doses of 500IU per kg body weight for female and 250 IU per kg body weight for males. Spawning was noted at 0430 hours on 13th March, 2010. The total eggs spawned were estimated as 2.1 million numbers. About 90 per cent fertilization was recorded (fertilized eggs were estimated as1.9 million).The eggs were collected by a 500 micron mesh cloth and stocked in incubation tanks with varying densities. The eggs hatched after 22 hours of incubation at a temperature range of 28-30 degree Celsius. The percentage of hatching was 80 per cent and the total number of newly hatched larvae was estimated as 1.5 million. The newly hatched larvae measured in total length from 2.2-2.7 mm each.The mouth opening was formed on 16th March, 2010 (on third day post-hatch).The mouth opening of the newly hatched larvae measured around 200 micron. The larvae were stocked in 15 FRP tanks of 5 tonne capacity each at an average density of 50000 larvae per tank for intensive larviculture. The remaining larvae were stocked in three 100 tonne cement tanks for extensive larviculture trials, The intensive larviculture tanks were provided with green water at a density of about 100000 cells per ml and also rotifers enriched with DHA SELCO at a density of 6-8 numbers per ml. In the extensive larviculture tanks green water along with rotifers is maintained. Good survival of larvae was observed.

Breeding of Marine Food Fishes in India

The honeycomb grouper, *Epinephelus merra* was attempted for breeding and seed production at Mandapam Regional Center of CMFRI. One broodstock tank of 5 tonne capacity was set up with undergravel filter. Six pre-adult fishes were stocked in the tank and fed *ad libitum* with fresh sardines. The fishes ranged in length from 20-36 cm and in weight from 100 to 650 g.

The fishes above 30 cm formed pair and natural spawning was obtained. During August-September 2005, seven spawnings were obtained. The periodicity of spawning ranged from 3 days to 12 days, but the interval in majority of spawning ranged between 3 to 4 days, The approximate number of eggs in different spawnings ranged from 11220 to 63020. The eggs hatched on the ame day of spawning. The average length of newly hatched larvae was 1.5 mm.

Larval rearing was conducted in 5 tonne FRP tank. Before the introduction of larvae, the tank was filled with filtered sea water and micro-algal culture was added to make the water green (green water technique). Calanoid copepods were introduced into the tank at an average concentration of 500 numbers per liter. The copepods were maintained in the tank in the multiplicative phase as was noted by the availability of egg bearing copepods, nauplii and copepodites.

About 2000 newly hatched larvae were introduced. Eighty percent mortality occurred during 3rd and 4th day. Thereafter the availability of sufficient copepod nauplii in the rearing tank was the key factor noted for the survival of the larvae up to two weeks. Whenever there was a decline in the availability of nauplii, mortality of the larvae was noted. Additional copepods were collected from the wild and added to the rearing tank to maintain the density of nauplii. The addition of rotifer to the rearing tank resulted in the blooming of rotifers with a consequent depletion of copepods in the rearing tank. This was found to increase the mortality of the larvae. Hence the maintenance of copepods in sufficient densities in the rearing tank was found to be the critical factor for the survival of the larvae. Afer two weeks, freshly hatched *Artemia* nauplii were also added to the rearing tank. From the25th day onwards, in addition to *Artemia* nauplii, adult *Artemia* was also applied as feed. The larvae started metamorphosing from 40th day onwards and all the larvae metamorphosed by the 60th day. A total of 33 number of young ones were produced in this trial. The young ones ranged in total length from 20-64 mm and the majority was in the length range 30-49 mm.

Epinephelus malabaricus

At Vizhijam, pre-adults of *E. malabaricus* weighing more than 1 kg and up to 2 kg were collected from April 2006 onwards and reared for development of parent fish stock. they were fed enriched diet and the required hormones for sex reversal also were administered to them. Male hormone was administered by incorporating through feed, twice a week from the first week of August. The dosage of hormones was at the rate of 3 mg per kg body weight of the fish. Eleven numbers of parent fish of *E. malabaricus* weighing from 2.85 to 5.45 kg were developed at the mariculture laboratory. Biopsy examination of the parent fish was carried out in January as well as during March, 2007. Two of the females were found to have the ova in the tertiary stage of vitellogenesis, measuring in size from 360 to 400 micron. Though there was no free flow of milt, sex inversion had taken place by the hormone application in the male parent fishes.

Siganus canaliculatus

At Cochin hatchery, four pairs of broodstock of the rabbitfish, *Siganus canaliculatus* were maintained in 5 tonne FRP tanks having *in situ* biological filters. Feeding of fishes was done with chopped fish meat, mussels, clam meat and fish eggs. Occasionally intertidal green seaweeds like *Ulva* were also given as feed. Water exchange in this tank was done 2-3 days before and after full/new moon days.

Spawning of the fishes was observed during November. Two spawnings were observed during this month (1st and 18th November, 2006). The interval between two spawnings was 18 days. Since the fertilized eggs were demersal and adhesive in nature, collection of eggs from the tanks was not possible. So the hatched out larvae were collected next day morning and transferred to larval rearing tanks.

The larvae collected from the tanks were stocked in one tonne FRP tank kept under roofing for further rearing. The larvae were stocked at a density of 5 larvae per liter. The size of the 2 day larvae was 2.87 mm. the mouth size of the larvae at that time was 100-125 micron. Rotifers at a density of 5-10 per ml were maintained as first feed. Every day morning the bottom of the tank was siphoned out and 5-10 per cent water was replaced with fresh sea water. The rotifer density was adjusted by adding fresh rotifers every day. To provide green water to the larvae and also as feed to rotifers, *Nanochloropsis* was added to the tanks every day morning and evening.

On 4th day a slight reduction in the number of larvae was observed. On 6th day the larvae had grown to a size of 3.25 mm and mouth size increased to 150-175 micron. The gut content analysis of the larvae revealed the presence of rotifers along with algae in the gut of the larvae. There was a gradual reduction in the number of larvae during succeeding days. Large reduction in the number of larvae was observed between 12th-15th day. At this time larvae reached the size of 5-6 mm and were actively feeding on rotifers. On 16th day, freshly hatched *Artemia* were added to the tanks as feed to the larvae. Between 20th and 25th day the larvae metamorphosed to juvenile fishes. A total of 35 juveniles were produced from this tank.

Breeding of Mullets

Excepting *Liza corsula*, which was observed to mature in Pulta Water Works settling tanks, West Bengal there is no report of other mullet species attaining full maturity in freshwater ponds. Pakrasi and Alikunhi (1952) observed of *Liza corsula* in fresh water rivers during June and July and reported that the fish appears capable of breeding both in brackish as well as in freshwater habitats and the breeding season seems to be a prolonged one extending from May to September.

Mugil cephalus is the most important cultivated mullet and has a wide distribution. The only means of obtaining the seed of this species is by collecting it from the sea coast, creeks, lagoons and adjoining areas. Besides the difficulties of collection and transport involved, the fry collected are usually a mixture of economic and uneconomic varieties. Attempts were made in few countries to induce them to breed by injection of hormones and raise the fry artificially.

Preliminary attempts were made in India to breed the two important economic species of Chilka lagoon, namely, *Mugil cephalus* and *Liza troschelii* by hormone treatment. Experiments were conducted in lagoon water and parent fishes were collected from the sea coast near the lagoon mouth. Injections of homoplastic pituitary gland extracts (8 mg to 16 mg per kg body weight of parent fishes) resulted in ovulation of eggs on several occasions in both the species. Both the species could be stripped successfully, but the eggs failed to develop in lagoon water. When sea water is used, normal development of eggs proceeded on several occasions and on one occasion a *Mugil cephalus* egg hatched out in 48 hours (22.5 to 23.5 degree Celsius) after fertilization.

The low salinity of lagoon water was found to be responsible for failure in getting fertilization of eggs in lagoon water. It was observed the *M. cephalus* sperms died immediately when brought in contact with lagoon water. *L. troschelli*, sperms, however, were active for about a minute. On the contrary, sperms of both the species were found to be active for about 10 minutes when mixed with sea water.

The results indicated that it may be possible to get successful spawning of these mullets if experiments were conducted in sea water with facilities for water circulation which appear essential for proper development of eggs and survival of hatchlings.

This work was continued and the success has been achieved by Odisha State Fisheries staff in 1963 in artificially fertilizing the stripped eggs of injected *L. troschelii* in sea water and 10000 spawn were produced which survived over a week.

Tang and his co-workers in Taiwan carried experiments during 1963 mullet fishing season and successfully bred *Mugil cephalus* by hormone injections (Tang, 1964). The parent fishes were collected from the sea during their spawning migration to the south-western coast of Taiwan. Parent fishes were held in special boxes in shallow sea water. The breeding technique followed was identical to that used for breeding of Chinese carps. Mullet pituitary extract in combination with Synahorin (a mixture of chorionic gonadotrophin (CG) and mammalian pituitary extract) was used. No ovulation occurred when1.0 to 3.0 pituitaries alone were injected or when the fish received combined extract of less than 2.0 pituitaries plus 40 rabbit units of Synahorin. Positive results were obtained when 2.0 fish pituitaries combined with 40 rabbit units of Synahorin were injected, which is considered to be the threshold dosage for successful spawning. Artificial fertilization of hand stripped eggs was made following the dry method. The developing eggs were hatched in aquaria with circulation of sea water and also under natural conditions near the sea coast. Hatching took place in 59 to 64 hours at a temperature 20 to 24.5 degree Celsius and in salinities of 24.39 to 35.29 ppt. Most of the larvae died within two days after hatching and none survived beyond 4 days.

High mortality of parent fishes after injection hampered the experiments. The authors believe that the response of these mullets to hormone injections could be greatly improved if the fish were kept in better condition at the time of experiment.

The preliminary success in India and Taiwan will go to a long way to develop a dependable technique for the spawning and successful rearing of the larvae and early fry of mullets.

Results of Hormone Treatment on Fish

Successful hormone treatment on fish and its use in pisciculture started in Brazil in the early thirties, when fish pituitary hormones were first used for inducing spawning in fish. This success has given a fresh stimulus to physiologists to conduct intensive research on the physiology of pituitary gland and its effect on the reproductive behaviour of fishes. Pickford and Atz (1957) in their book on "The physiology of pituitary gland of fishes" have summarized all available informations and discussed various aspects of the physiology of fish pituitary gland and its role on present day fish culture.

Although successful experiments have been conducted and techniques developed in a number of countries to breed fish by hormone injection, bio-assay methods have not been developed in any country other than Russia. Atz states, "Chemical standardization of fish gonadotrophin is of course a thing of the future, and the Russians are the only ones who have developed a method of biological assay".Although the Russians have found out a "Vy'un Unit" (Kazanski, 1949) for assaying fish gonadotrophin, they have yet to solve the problem of dosage of hormone.Gerbilskii (1965) has given an account of the technique of hormone treatment in fish culture followed in Russia. He states that the effectiveness of pituitary injections depends mainly on the coincidence of the time of treatment with the suitable period of the sexual cycle of fish. The study of fish ovogenesis is of great importance to "ensure effective application of pituitary injections and to forecast precisely enough the moment of mating". Sample of ovocytes are collected from injected sturgeon female every five hours during the period of maturation until the completion of ovulation. By making a thorough study of the complex processes, Deltaf and her collaborators have been able to recommend the proper time for obtaining eggs from the female. In Brazil, the techniques have not been standardized. In the absence of any accurate unit of measurement, Fontenele (1954) suggested a conventioal unit which is the amount of hormone obtained from a whole pituitary gland from a *Curimata* of well developed gonads. Clemens and Sneed (1962) attempted evaluation of the activity and relative potency of pituitaries of various species of fish collected during various months of the year and also determination of phylogenetic specificity, if any, in pituitary glands bio-assay methods. In India, however, although the pituitary hormones were not isolated and chemically standardized, fairly satisfactory results were obtained by precautions taken in selection of proper recipient fish (parent fishes) as well as the donar fish and also in preservation and storage of glands. The dosage given has been always in mg weight of the pituitary gland per kg of body weight of fish. Female breeders are also cathetered and eggs examined before final selection.

In general, pituitary glands from fully mature gravid fish are used in induced fish breeding experiments which are believed to be the most potent. Garbilskii (1940) observed relatively low potency of gonadotrophins in the pituitaries of spent fish and the same was observed by de Menzes (1945) in the case of immature fish. But Clemens and Sneed (1962) observed "Activity in relation to size, sex and degree of gonadal development of the donar fish does not appear as important as implied by some previous reports in the literature".As regards relative potency of the male and female pituitary glands, the results of research conducted in Brazil and also in CIFRI, Cuttack, India have shown that the pituitary glands from male and female fish are equally potent. Ramaswamy and Sundararaj (1957), however, observed that in the catfish, *Heteropneustis fossilis*, the pituitary gland of the male is not so potent as the female during breeding season. The results obtained by the said authors, however, are not strictly comparable as the relative potency was determined by injection of whole or part of a pituitary, and not according to the weight of gland or by any other bio-assay method. Besides there is a difference in size of the donar catfish as males are usually smaller than the females.

In preserving glands, both acetone drying and preservation with absolute alcohol have been found to be effective and are popular with the fish culturists all over the world. Neto and Rebeiro (1946) prepared fish pituitary extracts in glycerine and found that the solution retained its effectiveness for a year when kept sealed in airtight vials at room temperature. According to Menezes *et al.* (1945), glycerine preserved pituitary extracts are more uniform in potency than those in saline and distilled water. Ibrahim and Chaudhuri (1965) found glycerine method very effective and more convenient. Clemens and Sneed (1962) however, are of opinion that although glycerine extracts are easier to inject, their preparation is tedious. They also observed that the administration of more than 10 percent glycerine has harmful effect on recipient fish.

Phylogenetic specificity in fish pituitary hormones is a subject of controversy. Creaser and Gorbman (1936) believe that there is a specificity in hormone physiology and a qualitative species variation exists in gonadotrophic hormones. When the variation is sufficiently great between widely separated donar and recipient species, it leads to apparent ineffectiveness of the hormone. This view is also supported to some extent by Hoar (1957), who while reviewing the work of Witschi (1955) pointed out that the refractoriness of fish to mammalian gonadotrophins is probably due to different biochemical materials operating in groups as widely separated as mammals and fish. According to Pickford and Atz (1957) phylogenetic specificity does not appear to be significant in fish culture. Clemens and Sneed (1962) also observed that the specificity is so slight that it has no practical bearing in fish cultural practices. Chaudhuri (1960) reported successful spawning with heteroplastic pituitaries from fishes of the same family as well as from closely related families. Ramaswami and Sundararaj (1957) succeeded in spawning gravid catfish with homoplastic glands or with glands from genera of the same order but the fish became refractory when injected with pituitaries from fish belonging to different orders. However, the same authors reported successful spawning in the catfish by using frog pituitary hormones. Lin (1964) succeeded in breeding mud carp by injecting pituitaries from skipjack tuna.

These and other results achieved experimenting with mammalian pituitary hormones, sex hormones and various sex-steroids as reviewed by Pickford and Atz (1957) and Chaudhuri (1956) demonstrated the relative effectiveness of fish pituitary hormones on the spawning of fish, over those of mammalian origin. The fish pituitary hormone was, therefore, used in all the countries for spawning fish as a part of regular fish culture work.

Recent findings, however, show that the mammalian pituitary hormones in combination with the threshold dose of fish pituitary glands are able to precipitate spawning in fish. But out of all hormones of mammalian origin, the chorionic gonadotrophin has been found to be the most effective on fish. Hoar (1957) states " pregnancy urine and chorionic goadotrophin (CG) frequently give positive results but it is difficult to say which of the many factors present on these materials is active. According to Burrows (1949), the chorionic gonadotrophins behave primarily as a luteinizing hormone (LH) and since LH is believed to be responsible for the growth and maturation of gonads in fish, the effectiveness of CG on fish can probably be explained. Sneed and Clemens induced spawning in the channel catfish and obtained large scale production of fry by administering human chorionic gonadotrophin (HCG). Ramaswamy and Lakshman (1959) injected a pond catfish with 250 IU of chorionic gonadotrophin (known by the trade name of "Physex") and obtained ripening of eggs. But considering the high cost as compared to fish pituitary hormone, it was not tried in fish culture practices in India. In Russia, as early as 1947, HCH was used to diagnose pregnancy in women with the fish, loach as the test animal, but HCG was not immediately used in their regular breeding of sturgeons. With the increase in demand for pituitaries in recent years and difficulties in their mass procurement, the need of a suitable substitute was felt which could be produced commercially. "Choriogonin" (a trade name for CG) was first used in 1964 on loach with great success. Among other possible substitutes, several preparations, such as, Szhk, prepared from pregnant mare serum, "Estrovet" and "Hypophysin Forte" were tried by Bulgarian fish culturists on carps and trouts. These preparations reduce the duration of the spawning process.

Tang (1965), while summarizing the work of induced breeding of Chinese carp in Taiwan, stated that treatment with fish pituitaries in combination with chorionic gonadotrophins increases the effectiveness and better success be achieved. Synahorin was used with fish pituitaries and 78 percent positive response was obtained. He, however, does not mentioned the dose and number of systematic experiments were carried out with proper controls to arrive at the conclusion. Fish culturists are using different doses of Synahorin and fish pituitary glands and each hatchery has its own method.

Results of bio-assay experiment with trichloroacetic acid (TCA) and a few other chemicals on the glycoprotein content of catfish pituitary gland showed that both the glands immersed for 6 or 12 hours in 1.25 per cent TCA and the acid fluid in which they were immersed brought about spawning, but the glycoproteins are extracted and denatured in 12 to 24 hours. TCA of 2.5 per cent strength, however, did not give the desired results. TCA extract of pituitary glands was used in CIFRI, India for breeding of major carps. Pituitary glands immersed for six hours in 1.5 per cent TCA were successful in spawning in 11 out of 12 fish. The minimum effective dose was 14 mg per kg of the fish. Three sets of rohu when injected with 2.5 per cent TCA in which glands were immersed for six hours did not respond, but when the duration of immersing were reduced to three hours, another three sets could be spawned.

A number of techniques are followed by different workers which produce varying degree of success. The fish breeder should be able to select the best method suitable for the particular species of fish he breeds. There is, however, ample scope for improvement of the breeding technique by the use of proper bio-assay methods and by chemical standardization of the gonadotrophins.

Induced Breeding of Fish in North America

Carp (*Cyprinus carpio*) are not considered desirable fish in the United States and have limited sales at a relatively low price. However, when raised in ponds and fed well, the flavor compares favorably with that of other species, although numerous small bones limit the usefulness in the restaurant trade.

Parent fishes are carried through the winter in ponds stocked at the rate up to 5000 kg per hectare. They are fed 2 per cent of their body weight per day while the water temperature is above 19 degree Celsius. In March, small ponds, approximately 0.04 ha are filled with water from stream or pond that does not contain carp and are stocked with two pairs of parent fishes when the water temperature reaches 18 to 21 degree Celsius. Eggs are usually laid the next morning on submerged grass at the sides of the pond. Eggs hatch in tree to five days. Two to four weeks after hatching, carp are transferred to nursery ponds at the rate of 250000 per hectare. Ponds are fertilized regularly during summer.

Carp can be injected dried pituitary material at the rate of 2 to 3 mg per 0.45 kg of body weight and the sexes can be hand stripped 18 hours after the injections. The sexes are kept separate during the period. Eggs are hand stripped into a plastic container, and the sperms are added and mixed with the eggs. Small quantities of water are slowly added to mobilize the sperms, and just as the eggs begin to swell and adhere, they are dispersed with a nylon or plastic paint brush over the submerged spawning mat. The spawning mat then can be removed to nursery ponds. The method has the avantages of hormone spawning described elsewhere and can also be adapted to buffalow fish and gold fish.

The Israeli strain of common carp thrives under culture conditions and grows extremely fast but is more difficult to spawn than the common strain of carp.

Induced Breeding

The use of pituitary materials in North America to induce fish to spawn has many advantages in the culture of sport, food, bait and tropical fishes. Since fully ripe females of most species respond to pituitary injections by spawning within 12 to 20 hours, it is possible to predict rather accurately when spawning will occur. This predictability allows the fish culturists to plan his work according to a rigid schedule, since the injections to some degree by pass the environmental variable of temperature, light and rain. Both males and females respond positively to pituitary injections, so most species can be hand-stripped, a practice which offers additional advantages. Culture ponds can be stocked with eggs and fry of uniform size and age. Since the parent fishes are not in the pond with the offspring, the possibility of disease transmission from parent fishes to offspring is reduced, and the predation by the parent is eliminated. Hybrids can often be produced by hand stripping.

Wild fish that have not been domesticated may be spawned more easily by the pituitary method. Frequently, a more uniform spawning has been achieved by the injection of pituitary materials into ripe fish, which because of some environmental or physiological condition, would fail to spawn without such treatment.

Trial Considerations

Pituitary gland and its relation to the reproduction of fishes in nature and captivity together with fish gonadotrophins and their use in fish culture reveals :

1. Fish pituitaries contain the hormones necessary to precipitate spawning in ripe fish.
2. There is little, if any specificity in the hormones of fish, that is, the pituitaries of one family or species are usually active in unrelated species.
3. There are a number of examples in which gonadotrophic materials from other vertebrate classes were found to be active on fish.
4. Out-of-season spawning has been induced by the administration of gonadotrophins.
5. Fresh, acetone-dried, or frozen pituitaries, as well as glycerine extracts, saline suspensions and water extracts have similar activity.
6. Under most experimental conditions the injection of pituitaries does not decrease fertility.
7. Increased or decreased exposure to light (depending on species offish involved) may have the same effect on sexual development and spawning as pituitary injections.
8. There is little or no qualitative difference in the pituitaries of male and female donor fish, at least from the practical view point.
9. There may be seasonal difference in the biological effect produced by the pituitaries taken from fish prior to spawning and those taken immediately after spawning.

Since 1957, more information has been added, and some of the earlier ideas have been modified. A relatively quick and sensitive assay using fish has been developed to detect the principle in fish pituitaries that is responsible for ovulation in the female and seminal production in the male. Clemens and Johnson (1964) used this assay to re-examine the problem of specificity. These workers examined the response of gold fish to pituitary material from 20 species of vertebrates representing 17 genera, 13 families, 8 orders and 5 classes. They concluded that the magnitude of response roughly approximated the phylogenetic closeness of the donor to the recipient. If however, the donor and the recipient are rather distantly related, the dosage should be increased slightly.

Handling has been shown to have a direct effect on the gonadal processes of both male and female carp and gold fish, but the resulting stress is overcome by the injection of pituitary materials. This observation re-emphasises the value of hormones in spawning wild fish.

The importance of food on gonadal processes was demonstrated by Clemens and Grant (1964), when gold fish starved for five days required 16 days with food to regain their original gonadal condition. Further studies show that fish kept on a maintenance diet for 60 days temporarily loose their powers of gonadal development. That is, the gonads of such fish do not develop in the expected manner when the fish are placed on a growth diet. This observation suggests at least temporary impairment of gonadotrophic function of the pituitary.

Clemens and Grant (1964) showed that there was an optimum temperature with respect to dosage of hormones. Fish should be provided optimum temperature during both the gonadal development and spawning periods.

Long term growth of oocytes (from oogonia to ripe ova) was induced in gold fish by a bi-weekly injection of partially purified fish gonadotrophins with 6 to 12 times an average spawning dose. Thus genital development can be induced as well as spawning. The hastening of fish into spawning, the attainment of more spawns in a given interval of time, and in short a more regulated control of gonadal processes is suggested.

Since the pituitary size of small and big fish species varies directly with the size of the fish, it is important for injecting on gland basis to use a gland from a fish of similar size.

Although the differences in gonadotrophic activity may exist according to sex of the donor species or season collected, these differences appear to be insignificant in fish culture practices when pituitaries are injected on a weight basis.

Degree of Success

A review of the attempts to spawn fish by hormones reveals a number of techniques which produce varying degree of success. The reasons that a proper and proved technique may produce negative results lies to some extent with the qualifications of the investigators who is usually either an endocrinologist who know almost nothing about fish culture or a fish culturist who is rather un-informed about endocrinology. For this reason, only the positive results are regarded of induced spawning experiments as truly significant. Sometimes even the positive results need to be regarded with caution since fish can occasionally be chemically or physically shocked into spawning.

Clemens and Sneed's (1962) Method for Selection of Female Fish

The best method for selecting fish for advanced sexual development is to examine the fish carefully. The assessment of the females physiological receptivity to gonadotrophins should be based on a combination of the following characteristics;

1. Abdominal distention by the ovaries, preferably after the digestive tract has been emptied;
2. Condition of the genital opening;
3. Colour;

4. Degree of abdominal flaccidity; and

5. Behaviour

The assessment of the degree of sexual development of the male is not usually a major problem. During the spawning season, males of many species often show sexual dimorphism and any male in breeding colour is assumed to be ripe. If sexual dimorphism does not exist and it is difficult to determine ripe males from unripe females and immature individuals, the fish of undetermined sex can be stocked at a ratio of two or more per gravid female in an effort to ensure the presence of some mature males.

These fish should be similar in size to the selected gravid females. In those species that pair, the behaviour of the male frequently indicate whether or not he is a potential spawner. His positive or negative attitude towards the female or to other males often reflects his sexual development. A thorough knowledge of the natural spawning behaviour and conditions for the particular species makes it much easier to select fish that can be induced to spawn. Both males and females are usually injected for hand stripping procedure, but otherwise only the females receive gonadotrophins. For species that require more specific environment for spawning, such as, nest builders, it may be necessary to provide certain facilities, such as, vegetation, gravel, rocks cans or flowing waters. Stimuli from the environment may also play an important role in the proper metabolism and in the activity of the injected hormones.

Other important factors of general technique to be considered include the collection of pituitaries, dosage and handling. Pituitary may be collected from culled parent fishes or other fish and used in fresh condition. They may also be collected prior to the time they are to be used, dehydrated in acetone and stored in a dried condition. Fresh-frozen pituitaries are equally satisfactory. In using pituitaries very little or no attention is paid to the sex of donor fish or to the season at which they are collected. However, pituitaries of various species are usually kept separate.

It usually saves time and labor to give doses above the minimum required for spawning rather than to administer a sub-minimal dose. The evidence indicates that once the spawning dose is reached greater amounts of pituitary do not change the results.

Needle sizes from 19 to 21 are satisfactory and do not often clog. Luer-lock hypodermic syringes, which have a locked needle, will prevent the loss of materials if the needle clogs. Large fish are usually not anaesthetized. They are injected intra-peritoneally with pituitary doses of 2 mg or more of acetone dried whole ground pituitary per 0.45 kg of body weight with one ml or more of distilled water. Ten thousand units of penicillin may be used to combat infection of multiple injections are necessary.

Human chorionic gonadotrophin, obtained from urine of pregnant women, is commercially prepared and available at drug stores throughout the country. It has been found to be active in many species of fish, namely, channel catfish, gold fish, trout (*Salmo gairdneri*), striped bass, paddle fish, flat head catfish, blue catfish and white catfish. It can be obtained in liquid or powder form at a price comparable to fish pituitaries.

Suitable facilities and proper spawning places (such as well-aerated running water, aquaria, troughs, pens and ponds) must be provided, the type depending on the species to be spawned. If a question of water temperature arise, a temperature approximating that at which spawning occurs in nature is used. In such species as carp and gold fish where there is a danger that the eggs will be eaten, the fish can be made to spawn on mats which are moved periodically to hatching ponds or troughs. The egg masses of the catfishes are easily moved to mechanical hatching facilities immediately after spawning is complete.

The response of a particular species of fish to pituitary injections varies with the gonadal development of the fish. Very gravid females respond much more readily than the fish less well-developed. In other words, it is easy to induce spawning during the regular spawning season, and the fish often spawn with one injection, whereas fish injected a few weeks before their regular spawning season may require several injections over a period of hours or days in order to precipitate the spawning act. Usually several injections must be used to spawn fish past their normal spawning period, but spawning can not be obtained if degeneration of eggs has began. Usually delayed fish are refractory, but quite often normal eggs can be procured if several low-level dosages are given. Apparently the major requirement for successful induced spawning are fish that contain well-developed eggs and injected pituitaries which contain the components necessary to precipitate the spawning act or to produce ovolution. In many species, sperm is available in at least some males most months of the year. Immature fish respond abnormally to materials similar to those injected into mature fish.

Selection

Very little work has been done for selective breeding of warm water food fishes. The main difficulties lies being; (i) variation in spawning time; (ii) peculiarities of spawning behaviour; (iii) peculiarities in environmental requirements for spawning, particularly, space, visual and tactile stimuli; and (iv) uncertainity of regular spawning. Such problems should all be solved by the proper use of hormones as they have been in the case of channel catfish.

The goals of a good hatchery is required to be; (i) greater efficiency and economy of operation; (ii) greater egg production with a greater production of fish; (iii) greater resistance to disease and parasites; (iv) faster growth; (v) earlier sexual maturity. A successful program of selection can help in achieving these goals. Greater efficiency and economy, should result from greater production, less space requirement, fewer parent fish stock, lower mortality rates and shorter periods of maintaining those fish produced. Hormone injection and genetic selection will bring about year round production of any species, especially if light and temperature can be controlled.

Smith (1961) noted that in a relatively stable and protective environment, such as a hatchery, a new set of conditions is provided where artificial selection makes possible a more rapid genetic shift than is likely to occur in the natural state. Intensive and continuous artificial selection may therefore result in observable changes in a relatively short period of time.

The fact that hormones can be used to induce ovulation and seminal production has led the way to hand-fertilization technique, a technique for carp. The freezing and storage of fish semen will enhance this technique still further.

As regards to ovulate the various species of catfish and to handle the eggs, Duprea *et al.* (1965) reported that the female of a species to be crossed is injected with a spawning dose of HCG and placed with male of the same species in an aquarium. When the first signs of spawning behaviour are noticed, the female is removed and stripped after waiting a period of half an hour. This permits to more eggs to ovulate and at the same time they do not lose their viability, which occurs when catfish eggs remain an hour or more in ovarian fluid. Since male catfish can not be stripped, a male of another species must be sacrificed, the testis removed and macerated in an appropriate concentration of frog Ringers solution. The macerated testis are added to the eggs as they are stripped into a small amount of water. The eggs are swirled over a piece of polyethylene plastic and allowd to settle to the bottom where they adhere. With this technique they have obtained hybrid crosses from both male and female combinations of the following channel catfish (*Ictalurus punctatus*), white catfish (*Ictalurus catus*), blue catfish (*Ictalurus furcatus*), flat head catfish (*Pylodictis olivaris*), black bullhead catfish (*Ictalurus melas*), yellow bullhead catfish (*Ictalurus natalis*) and brown bullhead catfish (*Ictalurus nebulosus*). Crosses in which female flathead were used all proved unsuccessful. The embryos died before hatching or shortly afterwards. When the male flathead was used in hybridization, it proved successful in four other species that were tried.

In this study 15 hybrids have been produced from seven parent species. Each hybrid is studied and ovulated for use in fish culture practices.

Hand fertilization methods make it possible to divide the semen and ova into a number of portions and to obtain several crosses from individual fish at one time. Facilities and suitable marking techniques are needed to promote these studies. This technique will be of great advantage to the geneticist because they give him the means to speed up the study of fish genetics.

Three species of buffalo fish have been hybridized by Guidice (1964) with hormones and hand stripping methods. Only black X big mouth spawn naturally without hormones in the pond. The hybrids grow faster than parent species, as much as 33 per cent by weight for black X big mouth.

Chapter 4

Breeding of Grass Carp and Silver Carp

Biological Characteristics of Phytophagous Fishes

Of all the phytophagous fishes, only the grass carp, *Ctenpharyngodon idella* and the silver carp, *Hypophthalmichthys molitrix* are significant in commercial fisheries. Both these species are known to occur in the river.

C. idella is a quick growing fish attaining more than 30 kg in body weight. For a month and a half its fry feed on animal food, mostly zooplankton. As soon as fingerlings are 3 cm long, they become phytophagous. *C. idella* lives on higher aquatic plants including soft submerged vegetation, young seedlings of some submerged water plants and meadow plants submerged in high floods. Besides higher plants, the grass carp willingly feeds on filamentous algae.

H. molitrix is another rapidly growing fish which reaches the weight of more than 16 kg. Unlike grass carp, this species feeds on lower aquatic plant, that is, microscopic planktonic algae filtered by the gill apparatus. The gill rakers of the silver carp make a fine-meshed sieve which serves to collect the phytoplankton entering the oral cavity.

The fry of silver darp at first feed on zooplankton, and only when the fingerlings reach the length of 1.2 to 1.7 cm, they become plant feeders. Adult specimens eat phytoplankton almost exclusively. However, when phytoplankton is not available, they may feed on detritus.

For feeding, phytophagous fish enter numerous flood-plain lakes in the basins of Amur and some Chinese rivers. In autumn, during the abatement of the river and fall of temperature, both the grass and silver carps migrate towards the river bed where they hibernate and usually ceases feeding.

In spring, mature fish migrate to areas with swift currents for breeding. Spawning begins when the water temperature is about 20 degree Celsius. Besides the temperature factor, spawning in both species is influenced by fluctuations in river level. Japanese investigators think that the onset of spawning is determined by water level changes, and that the temperature factor plays subordinate role in the process (Sudzuki *et al.*, 1962, 1963).

In the rivers of South China, both species attain sexual maturity at the age of three or four years. Further to the north plant-eating fish mature later. Thus in the basin of the Yangize river the silver carp mature for the first time at the age of four years and the grass carp at the age of five or six years. In the Amur both the species mature even later (Makseva, 1963).

In the Chinese rivers spawning of both the species takes place at the same time, but only a part of the stock, spawns at a time. This is why the spawning period is rather extended, lasting from the middle of April to August. In the Amur river, according to Makeeva (1963), specimens occur with asynchronous vitillo-genesis, indicating that the same female may spawn twice during one growing season.

An unfertilized egg is on average 1.16 mm in diameter, but on swelling, the initial volume increases from 45 to 100 times. The swelling capacity is also retained in unfertilized eggs. Depending on the temperature, the incubation period lasts for 20 to 40 hours. Hatched larvae are at first carried passively downstream and afterwards concentrate along the shoreline. For feeding they migrate to the system of flood-plain lakes.

Manipulated Breeding

Rapid growth and the habit of feeding on macrophytes and phytoplankton were the reasons for breeding the grass carps and silver carps. Due to intensive development of culture fisheries, required quantity of stocking material is not available from natural water bodies. Therefore artificial breeding of these species was considered to be the primary task of fish breeders.

In the Soviet Union, the breeding of grass carp and silver carp was tried in nursery ponds. Because of the absence of an established technique for artificial breeding, large quantity of fry were imported from the Far East and China to the European part of USSR. However, together with carp fry there came some undesirable trash fish, and even some dangerous parasites. A very complicate system of quarantine and the dependence of fishery managements on the yield of larvae of these fishes from natural waters greatly handicapped their culture.

Numerous attempts to induce breeding in special ponds imitating natural spawning conditions failed. Later work concentrated on the stimulation of maturing gonads with hormone preparation. In 1959, Soin took part in the work carried out on the Sungari River in China to obtain sexual products of the silver carp. There he obtained commercial quantities of the silver carp larvae from spawners which had reached maturity in natural water bodies. Despite this success, the use of hypophysial gonadotrophic hormone for breeding these fishes did not yield encouraging results. Very frequently the injected spawners did not attain sexual maturity, or the eggs fail to develop.

The spawning stocks of grass and silver carps included specimens which are not equally ready for spawning. Fractional injections of hormone might be effective on not-quite-mature females. Kasanakii (1960) had successfully applied fractional injections in his work with sevruga (*Acipenser stellatus*).

In China, in order to obtain eggs from phytophagous fishes, very large doses of hormone is used. Although large dosages of gonadotrophic hormones induced ovulation, they greatly affected the quality of eggs and these failed to ripen.

For artificial breeding of grass and silver carps, the dosage of gonadotrophic hormone should be chosen according to whether injections are made at the onset or by the end of spawning season.

Technique of Breeding

The grass and silver carps reared in stagnant ponds attain sexual maturity at the age of at least five years in the southern region of USSR. Work on obtaining eggs should be started not earlier than when the water temperature is 20 degree Celsius. Eggs and milt are obtained by means of hypophysial injections. For this purpose females of both the species are treated with fractional injections of wild carp hypophysial suspension. Males are treated with single dose injections. The purpose of the first injection is to stimulate preovulation nuclear changes in the ovocytes and is therefore only a stimulating injection. Nuclear rearrangements are completed 12 to 20 hours after this injection. That is why the second injection which must cause ovulation is made 24 hours after stimulation. The second injection is called resolving and its dosage is 7 to 10 times as high as that of stimulating one. 20 to 35 mg of hypophysial substance is injected. Depending on the on temperature, ovulation appears to be completed 8 to 13 hours after the injection.

Eggs are fertilized by milt not diluted in water (1-1.5 cubic centimeter milt per liter of eggs). After mixing the eggs with the milt, water is added. Then the mixture is allowed to stand for 10 to 15 minutes, and during this time it is shaken slightly. After this it is placed in a Weiss apparatus (8 liters) with a capacity for 50000 to 70000 eggs. A water current of 0.6-0.7 liter per minute is required for a standard apparatus.

Hatched larvae are very agile and therefore, they are carried out of the apparatus by the stream of water into special larvae traps of fine gauze netting (No 15-19). There they are scooped out together with water and put into stews made of the same fine gauze netting and installed in ponds. The oxygen concentration of the water in the stew should not be less than 4 mg/liter.

Stews must be protected against small trash fish, for the latter can suck the larvae out through the walls of the stew. On the fourth or fifth day the larvae begin to feed on the fauna around and swim freely in the depth of water. Now they can be released into ponds for rearing.

Larvae are transported in special plastic bags filled with oxygen. One end of the bag is sealed off, while the other end is hermetically closed with a clamp. The ratio of water and oxygen in the bag is 1:1 and stock density 2000-4000 per liter.

Chapter 5

Breeding of Sea Bass

Biology

Sea bass, *Lates calcarifer*, is a percoid fish belonging to the family Centropomidae, Order Perciformes. The eggs are spawned and fertilized in the sea and the larvae enter brackish water swamps and mangrove areas where they thrive on the abundant food found in these nursery grounds. Sea bass also inhabits estuaries, rivers and lakes and returns to marine waters to spawn, thus completing a life cycle spent in both freshwater and seawater. In fact sea bass is euryhaline.

Studies showed that juveniles mature initially as males, after 3-4 years but invert to females on the 6th year. However, not all males become females because primary females do occur. Sea bass, therefore, is a protandrous hermaphrodite. In captivity, sea bass can be spawned quite readily with or without the use of spawning agents.

Although there are no documented evidence in the Philippines of sea bass undergoing gonadal maturation and spawning in the wild, the onset of their natural breeding season is indicated by the appearance of sea bass fry in milkfish fry collections. The breeding season coincides with the monsoon months from late June until late October. Because sea bass is an incidental species in most fry collections, its collection from natural fry grounds is not reliable, largely inefficient, and tedious.

In general, sea bass is an opportunistic predator throughout its life cycle. Fish less than 4 cm feed on "microcrustacea" almost exclusively; 30-cm fish have diets of "macrocrustacea" and fish; larger individuals predominantly prey on fishes. In captivity, sea bass accepts pelleted rations.

Sea bass growth rate varies depending on culture conditions, but is generally high. Sea bass fed trash fish in cages grow from an average of 22 g to 573 g in 7 months, and in ponds from 7.8 g to 369 g after 7.5 months.

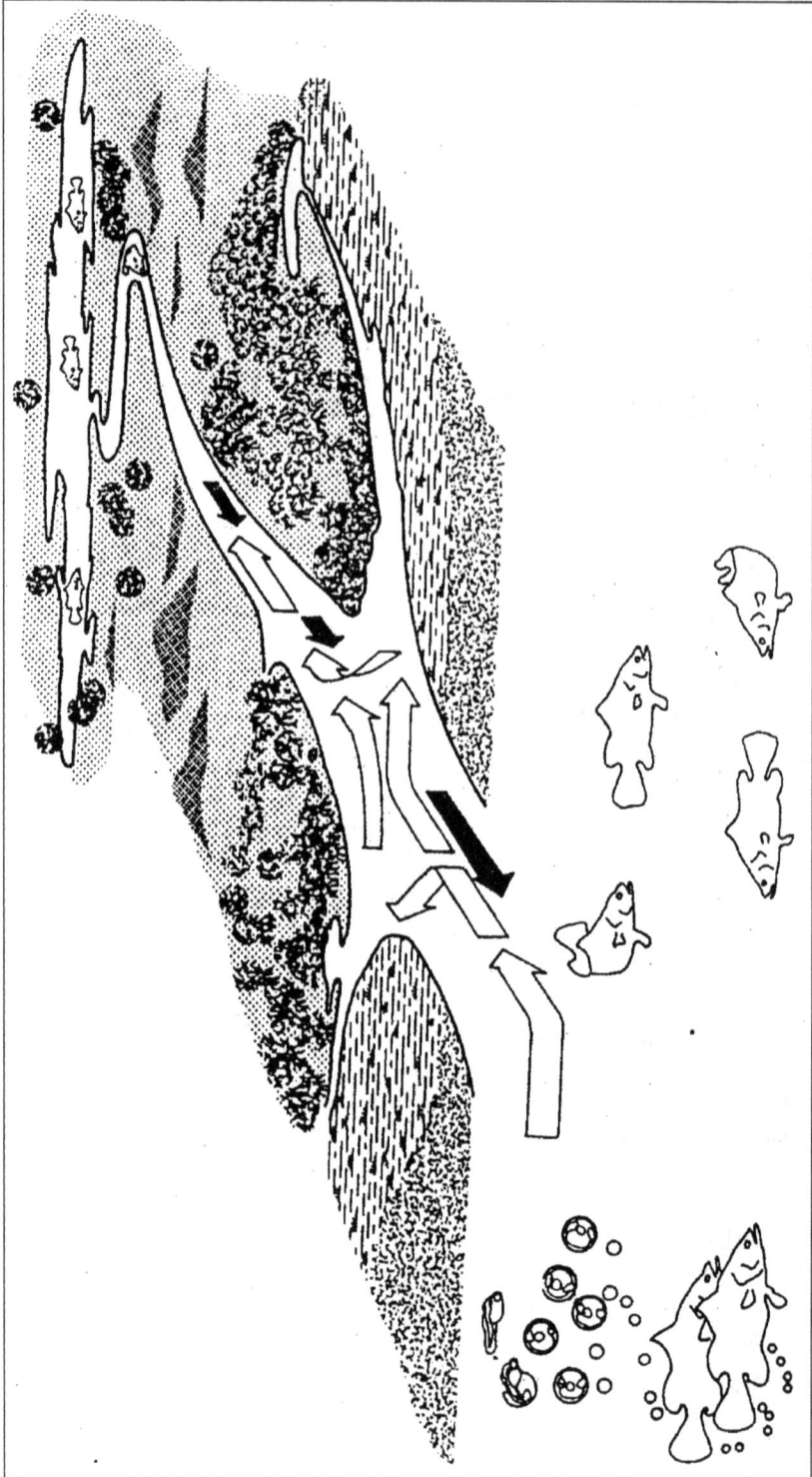

Figure 5.1: Life Cycle of Sea Bass including Migration Pattern from the Open Sea to Estuaries, to Freshwater Areas and Back.

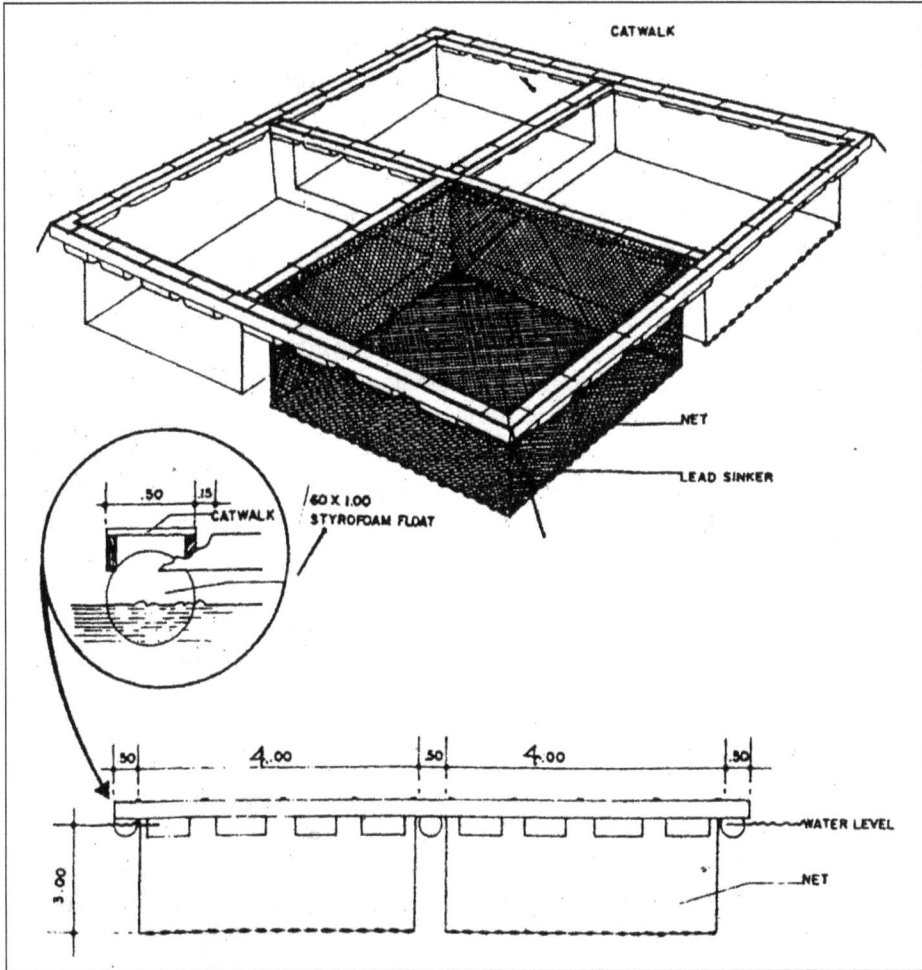

Figure 5.2: AQD-Designed Floating Rearing Cage for Sea Bass Broodstock. Material of wooden or bamboo frames, polyethylene net (5.2 cm stretched mesh), and styrofoam buoys (After Marte *et al.*, *SEAFDEC AQD & IDRC Tech. Reptr.* 11, p. 10. 1984).

Objective

Induced spawning and operation of a sea bass hatchery is aimed at maximizing economic return. For this reason, the economic as well as the biological factors involved in it must be understood. There are various ways of evaluating these factors but for practical purposes, it is sufficient to keep in mind the following points.

1. The market for which the product (fry) is intended must be identified and carefully assessed.

2. The cost of production under local economic conditions should be thoroughly appraised so that resources can be used efficiently. Since these considerations make use of biological phenomena as basis for production,

it is necessary to understand the biology of sea bass and the environmental processes occurring in the hatchery. Profit or loss largely depends on being able to ensure the maintenance of conditions conducive for larval development, growth and survival. In practice, the generalizations listed below has to be borne in mind.

3. Developing eggs and larvae are the most vulnerable stages in the life history of the fish and therefore require proper care, good quality water, and sufficient supply of suitable food.

4. " Good quality" water is a relative term; it refers to water that can support and maintain environmental requirements and sanitary standards necessary to sustain larval growth. Quality refers to the suitable temperature and salinity, sufficient supply of oxygen, and minimal concentration of ammonia, other toxic substances, and pathogens.

5. The effect of temperature on larvae is largely biochemical and is reflected on growth performance. Extremely high or low temperature cause death. Sudden temperature changes cause stress and stressed larvae give unsatisfactory return.

6. Changes in salinity influence the water balance of the larvae. In waters with high salt content, water is drawn out from the larva; in waters with low salt, the larva draws in water- both causing stress and eventual death.

7. Oxygen is necessary for almost all organisms to sustain life, oxygen must be available at all times especially when the hatchery carries heavy biological load.

8. Ammonia is the end product of protein breakdown among organisms, ammonia retards growth and causes stress and eventual death. Therefore, its content in the water must be minimized.

9. Toxic substances occur in the form of metabolites produced by the growing and feeding larvae, inorganic and organic wastes from sewage disposal systems inadvertently taken in with hatchery water, chemicals accidentally dumped in water and others. Such substances have adverse effects on the larvae, hence water must be free from these substances.

10. Pathogens occur anywhere in nature, hence can not be totally eliminated. The water used must not contain high numbers of pathogens. As the adage goes, an ounce of prevention is always better than a pound of cure.

Suitable Site for a Sea Bass Hatchery

The major criterion in selecting a site for the construction of a sea bass hatchery is the ecological requirement for spawning and egg and larval development. A suitable site, therefore, should be an area where access to a sufficient supply of pollution free sea water is possible. The salinity should range from 25 to 32 ppt. For practical reasons, the hatchery should also be near the source of parent fishes; however, a hatchery may have its own stock of parent fishes facilities.

The site should be spacious, gently sloping, and protected from strong winds and wave action. A gently sloping area gives the advantage of using gravity to distribute water from an elevated storage facility, and affords protection from water inundation or flooding during the rainy season. The area should also be located where natural protection from adverse weather conditions (storms) is possible.

The seawater source should have minimal siltation problems and should be far from mouths of rivers as freshwater runoffs will cause high turbidity and drastic fluctuations in water salinity. The site should also have an ample supply of fresh water. Freshwater is necessary when water of low salinity may be required. It is also necessary for cleaning and washing hatchery equipment and facilities.

The site should also have access to a reliable source of electric power and to transportation and communication facilities.

Design

Once a suitable site has been selected, the design of a hatchery must be considered. It is important to have in mind the production target and the available financial resources as these two factors determine the size of the hatchery to be constructed.

Tank Systems

Three major kinds of tank systems are needed in a sea bass hatchery, namely, parent fish stock, larval rearing and natural food tanks. Figures 5.3 and 5.4 shows a sample layout of a sea bass hatchery with parent fish stock facilities. The design may be modified according to available capitalization and the physical characteristics of the site. With certain modifications, most prawn hatcheries may be converted into sea bass hatcheries.

Parent Fish Holding Structures

Sea bass spawners are maintained in these structures which are installed either on land or in open waters. Land-based structures include circular or rectangular concrete tanks. The tanks are usually big and have water holding capacities of more than 50 tons.

Larval Rearing Tanks

Sea bass larvae are reared in concrete tanks (such as those used by SEAFDEC) or in circular canvas tanks, such as those used for prawns by private operators. These tanks are provided with roofings to protect the larvae from direct sunlight. Circular tanks have conical bottom and a centrally located drainage pipe connected to the canal. This design facilitates cleaning and harvesting.

Rectangular concrete tanks may also be used. These have a flat bottom and the drainage is usually located opposite the inlet pipe. There is no functional difference between circular and rectangular tanks, but the latter has practical disadvantages which include difficulty of siphoning dirt that accumulates in tank corners and

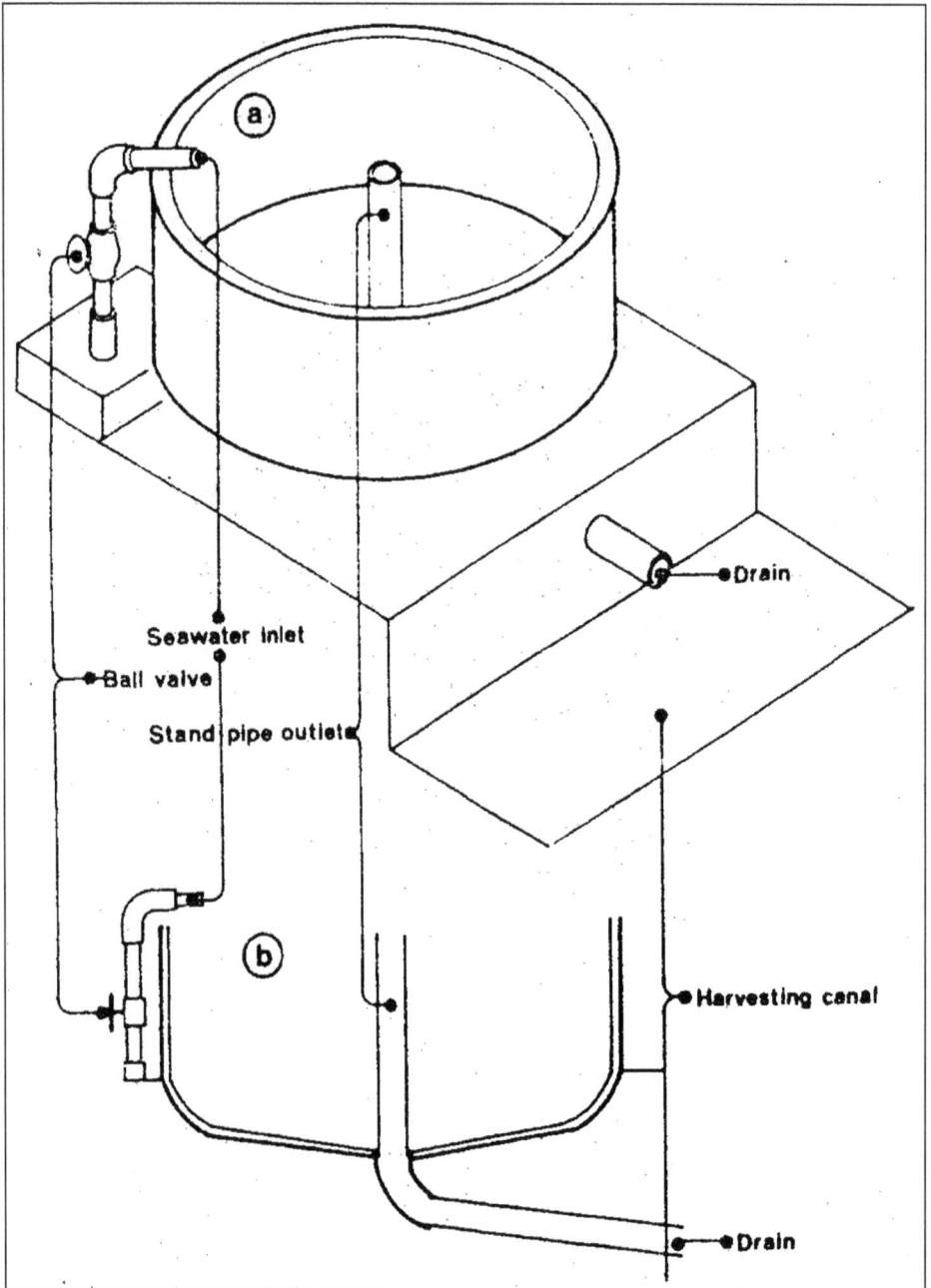

Figure 5.3: Circular Tank: (a) Isometric View; (b) Cross-sectional view.

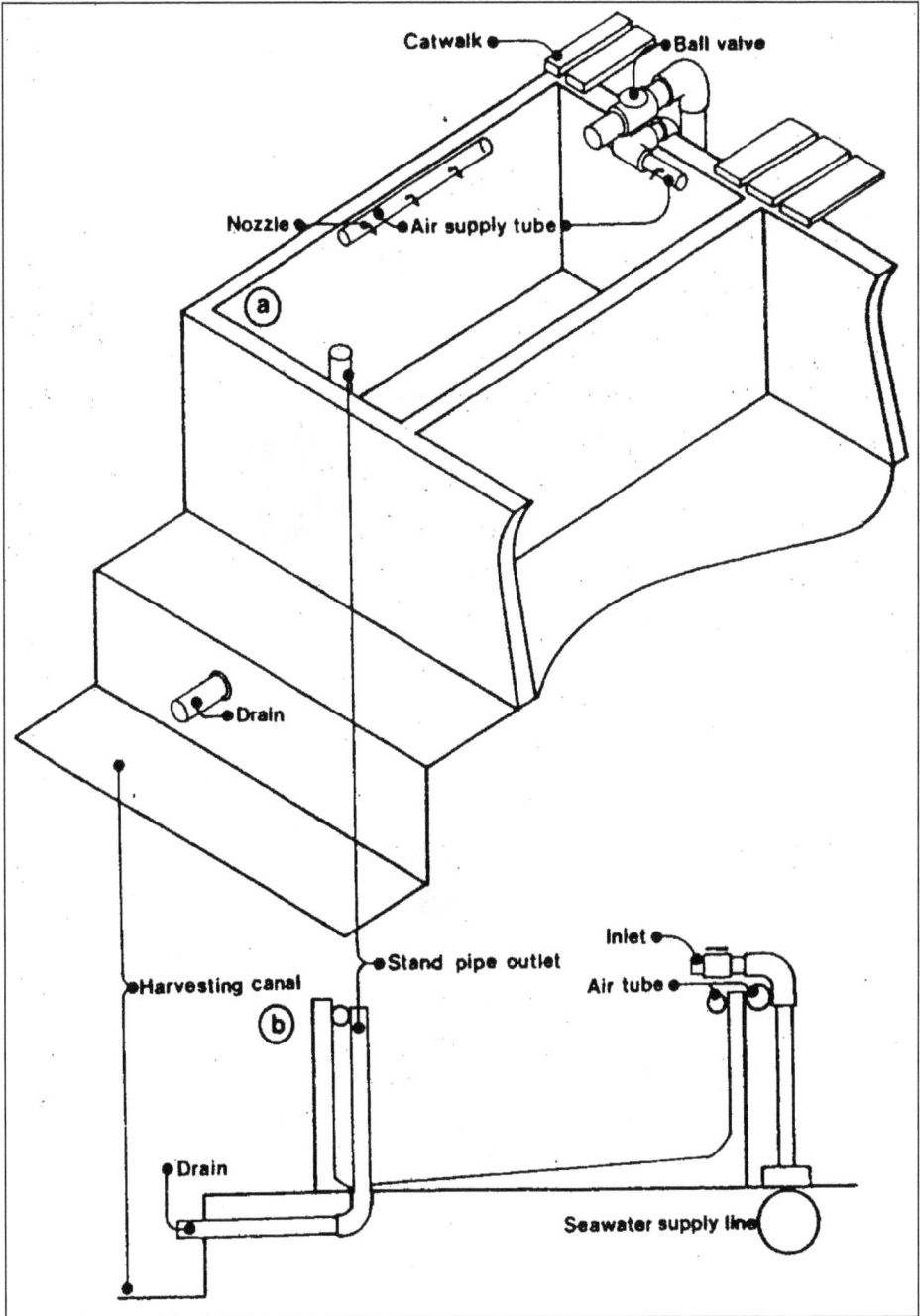

Figure 5.4: Rectangular Tank: (a) Isometric View; (b) Cross-sectional view.

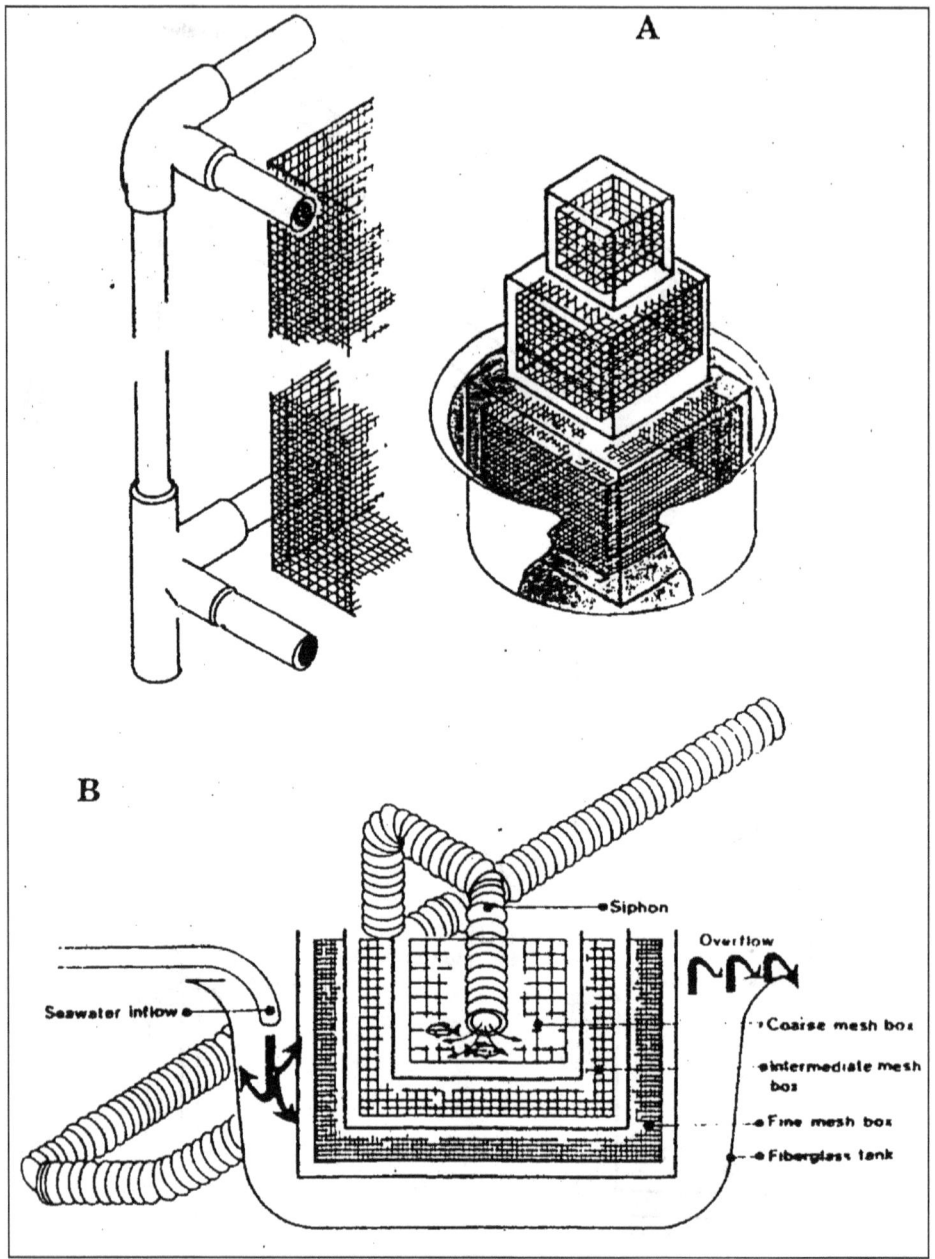

Figure 5.5: Sorter Boxes for Grading Sea Bass Fry; (A) Details and arrangement of boxes: (B) Cross-sectional view

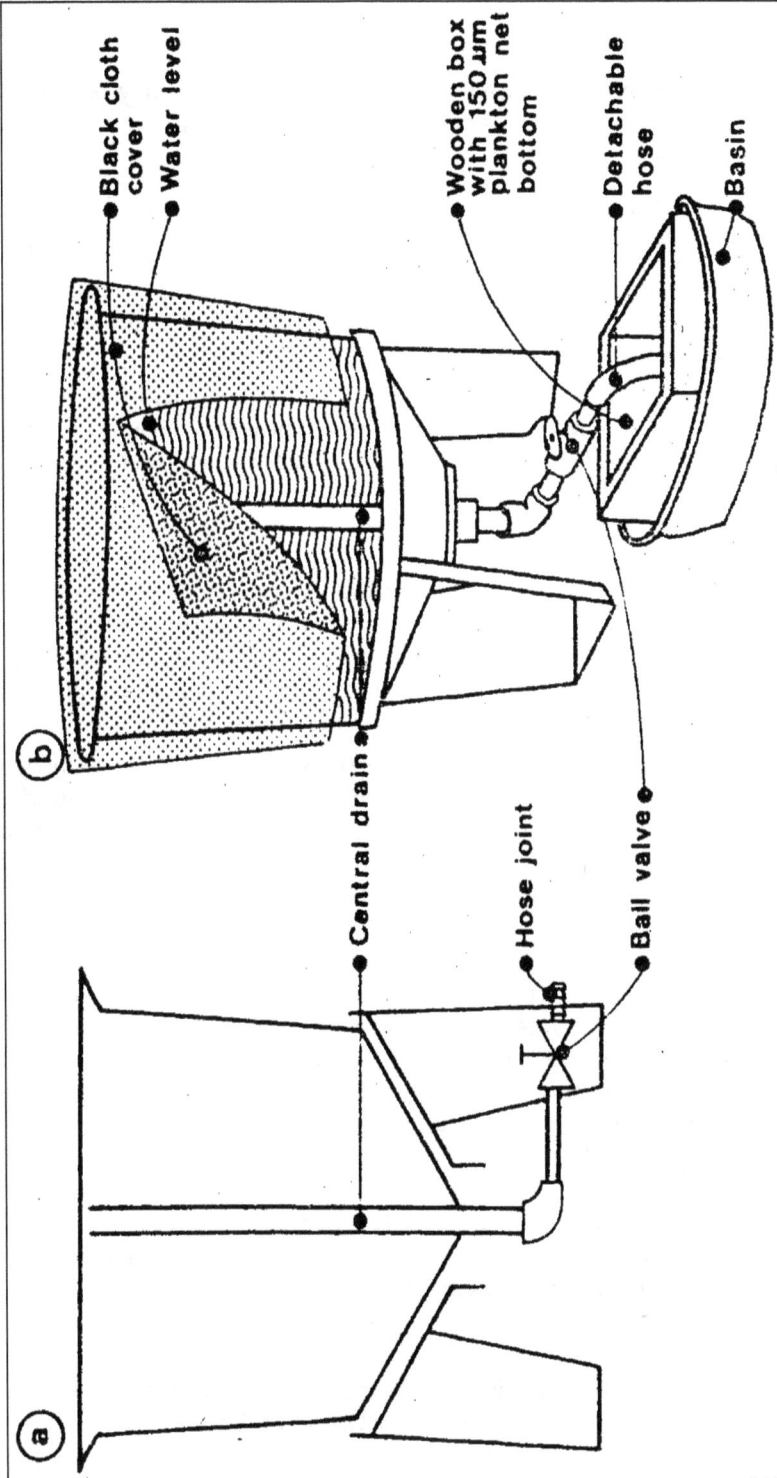

Figure 5.6: Artemia Incubation Tank Used at SEAFDEC-AQD; (a) Cross-sectional view: (b) Inside view.

Labels: Black cloth cover, Water level, Wooden box with 150 μm plankton net bottom, Detachable hose, Basin, Central drain, Hose joint, Ball valve formation of "dead corners" or portions of the water column in the tank where there is no water circulation

Canvas tanks can also be used for larval rearing. These may be cheaper to construct but do not last as long as concrete tanks. In the long run, concrete tanks will turn out cheaper.

Natural Food Tanks

Tanks are required for the culture of live food organisms, namely, *Chlorella, brachionus*. The tonnage ratio of *Chlorella* to *Brachionus* tanks is about 1.5:1, that is, for every ton of *Brachionus* tank, 1.5 tons of *Chlorella* is needed. The ratio of *Brachionus* to larval rearing tank is 1:0.5.

Circular canvas or rectangular concrete tanks with capacities of more than 10 tons can be used for the culture of algal food. These tanks should be situated outdoors to enhance propagation of *Chlorella* which needs sunlight.

Seawater Supply System

The seawater supply should be clean and free from pollutants. If the water is relatively clear, it can be pumped directly to an elevated filter tank, stored in the reservoir, and then distributed by gravity to the different culture tanks. During heavy rains when the seawater may become turbid, water may be pumped into a sedimentation tank to allow suspended solids to settle. Only the upper layer of clear water is pumped into the filter tank.

The pumping capacity of the marine pump needed in the hatchery will depend on the water volume requirement, pumping time, and total head. Total head is the difference in elevation between the surface of the source of water and the point of discharge. The daily water volume requirement can be calculated from the total volume of the tanks and the rate of water exchange.

Freshwater Supply System

In some instances, salinity lower than what is normal of sea water (35 ppt) is necessary, in which case freshwater is added to the sea water to bring down the salinity to the desired level. Also routine hatchery activities like cleaning and washing of tanks, basins, filter bags and other hatchery equipment require freshwater. The hatchery therefore, should have ample supply of freshwater

Aeration System

A roots blower is commonly used for this purpose. Water depth, number of aeration outlets, and cross-sectional area of the outlets have to be considered in determining the required blower capacity.

Other Facilities

In addition to the basic facilities and life support systems, the hatchery should also have a small laboratory, where a microscope and other laboratory equipment may be kept, and a store room as well.

Stock of Parent Fishes

An important aspect in the production of sea bass fry in a hatchery is the availability egg-bearing parent fishes. Although fry can be reared in captivity until they become fully ripe, a fish farmer often desires to produce sea bass eggs on demand and not to depend entirely on natural sources. Recent research results in SEAFDEC/AQD have become the basis for developing the following techniques to control the reproductive cycle of sea bass.

Source of Parent Fishes

Adult sea bass spawners may be obtained from natural spawning grounds or from brood stock farm. Wild spawners weighing 2 to 8 kg each are caught by gill nets, hook-and-line and fish traps, often near river mouths during the months of June to October. They must be wound and disease-free, with no missing body parts and strong and active upon capture. Ripeness of the reproductive organs (that is, gonads) can be checked as follows:

1. Fish is transferred into a shallow tank filled with sea water containing 250 ppm (0.25 ml per liter of sea water) of anaesthesia. Ethyleneglycol monophenylether (or 2-phenoxyethanol available from Merck, P.O.B. 4119, Darmstadt, Federal Republic of Germany) is a common fish anaesthetic.

2. Anaesthetized fish may be turned over on its back.

3. The abdomen may be massaged gently following a head to tail direction. A milky white substance of medium-thick consistency extruded out of the urogenital opening indicates the presence of milt among sexually ripe male spawners

4. When no milt is extruded after repeated massage of the abdomen, the tapered end of a polyethylene cannula (Clay Adams POE 100, inner diameter = 0.86 mm, outer diameter = 1.52 mm, available from Becton, Dickinson and Company, Parsippany, New Jersey 07054, USA), is gently inserted 10 cm into the urinogenital opening of the fish. The other end of the cannula is then gently aspirated by mouth as the inserted end is carefully withdrawn from the fish.

5. The contents of the cannula is inspected. A milky substance indicates milt, whereas tiny spherical bodies are eggs and the spawner is female.

6. Eggs may be blown out into a small vial containing 5 per cent buffered formalin solution as fixative.

7. A few eggs may be transferred on to a glass slide and the diameter of 30 eggs may be measured with a calibrated microscope. The average diameter of the egg may be calculated. A female having an average egg diameter of at least 0.40 mm is sexually ripe.

Wild and sexually immature adults can be reared in captivity until the gonads ripen. Parent fishes may also be obtained after 3-5 years of rearing fry and juveniles in sea cages, tanks or earthen ponds.

Breeding Techniques

Reproduction in all fishes is controlled by the hypothalamo-pituitary-gonad axis. The hypothalamus is a discrete region of the brain whereas the pituitary is a small gland located at the base of the brain. In response to various environmental stimuli, the chemical substances (hormones) secreted by the axis directly influence the onset of sexual maturation resulting in the shedding of ripe eggs and sperm. The sequential mechanism by which these hormones act on the gonads makes it possible to intervene in the process of sexual maturation and spawning. This is commonly done by the exogenous application of substances which mimic the actions of endogenously secreted hormones on the axis. Depending on the type of hormone applied to the fish, it acts mainly at the level of the hypothalamus, pituitary, or gonad.

Successful attempts to manipulate the reproductive cycle in sea bass have employed three hormones, namely, human chorionic gonadotrophin (HCG), luteinizing hormone-releasing hormone analogue (LHRHa) and 17 alphamethyltestosterone (MT). These hormones induce the gonads to mature earlier than is normal during the natural breeding season or trigger sexually mature fish to spawn. The following technique highlights on the use of these hormones.

Hormonal Induction of Sexual Maturation

This technique involves implantation of two hormones, LHRHa and MT, incorporated and pelleted in a matrix of cholesterol powder. Starting in late February, monthly implantation of these hormones at a dose of 0.1 mg per kg body weight results in gonadal maturation of male and female sea bass in floating net cages in May or two months earlier than the known peak breeding period in the wild. The procedure for hormone implantation is given below.

1. Condition healthy adult parent fishes (more than 4 years old) in floating net cages for at least 2-3 months.
2. Weigh each fish several days before implantation and prepare hormone pellets as described below.
3. Lift the stocking cage and gently scoop out fish one at a time.
4. Transfer fish into a shallow tank with 300 ppm of anaesthetic.
5. Turn the anaesthetized fish on its back with a pair of tweezers, pul out several scales at a point 7-8 cm fron the anus to expose the flesh of the fish.
6. Carefully make a short (0.5 cm long) but deep incision on the exposed flesh until the tissue lining the body cavity is punctured. No internal organ must be injured during the incision procedure.
7. Fit a metal trochar into an 8-cm long plastic drinking straw and carefully insert both into the incision. The trochar implanter and plastic straw guide will loosen when they reach the hollow body cavity.
8. Withdraw the metal trochar implanter leaving about 3 cm of the plastic straw guide protruding from the incised wound.

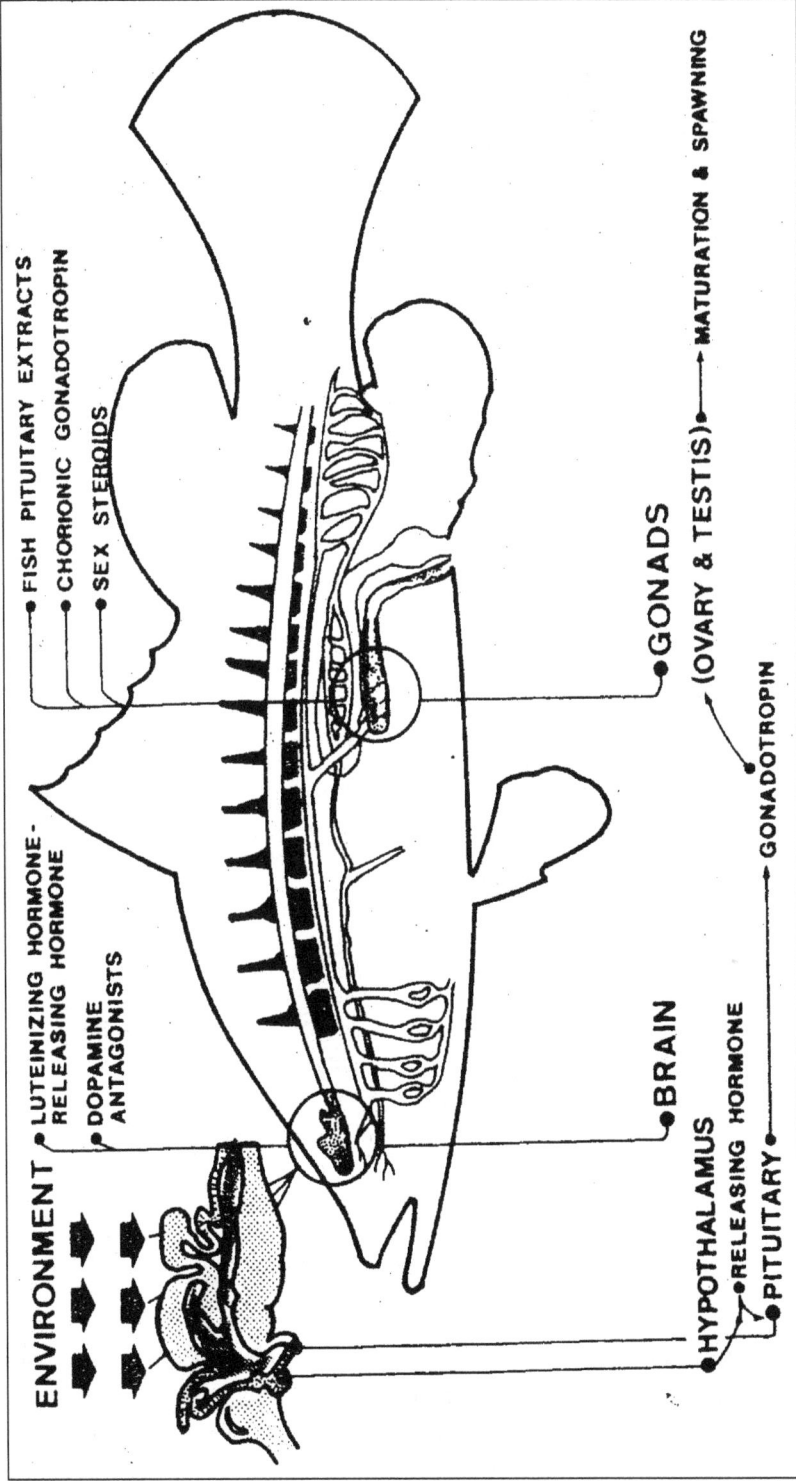

Figure 5.7: Levels of Exogenous Intervention in the Hypothalamus-Pituitary-Gonad Axis Leading to Gonadal Maturation and Spawning.

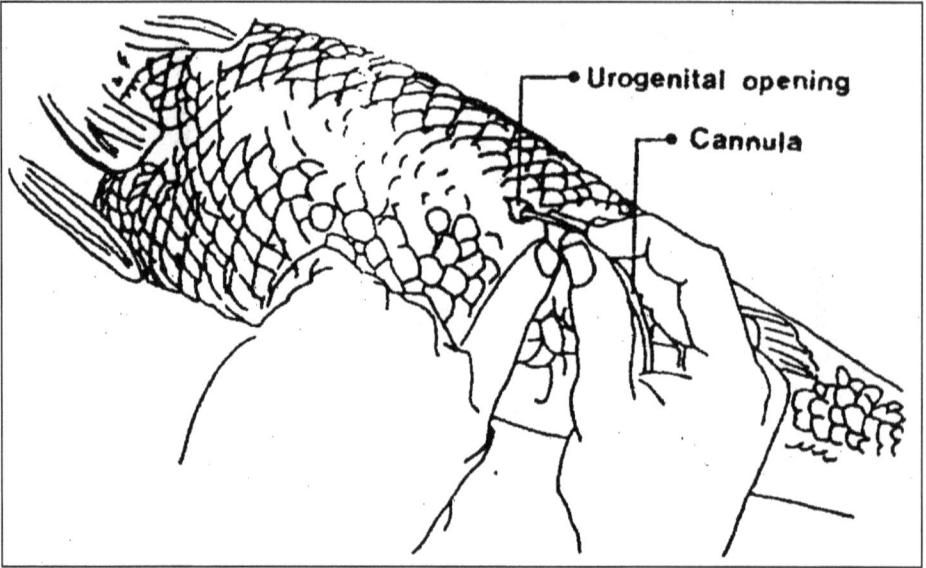

Figure 5.8: Sampling for Sex Determination. A cannula is inserted into the Urogenital Opening to Obtain Egg or Sperm Sample.

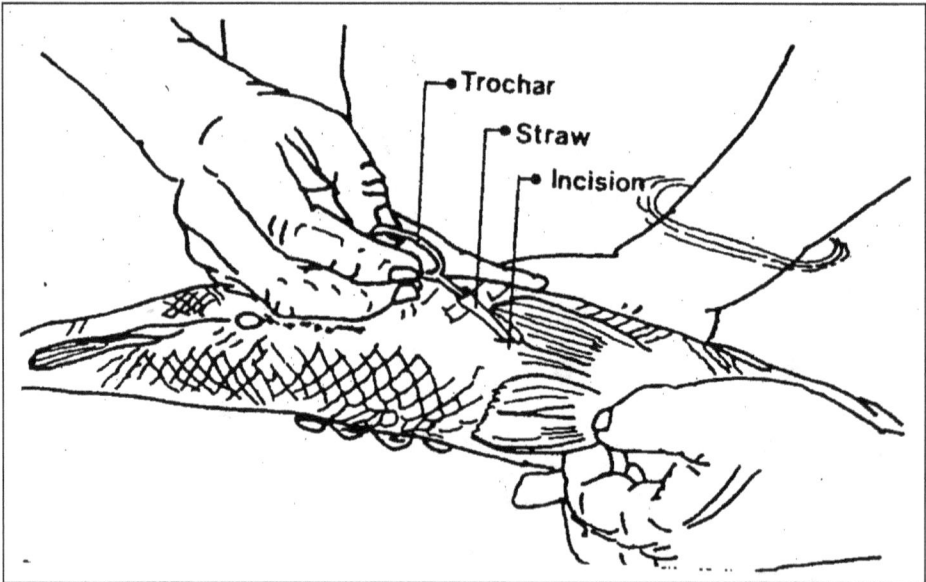

Figure 5.9: Implantation of Hormone Pellet Using a Metal Trochar and Straw Guide.

9. Drop the hormone pellets into the plastic straw guide and replace the trochar implanter to push the pellets into the body cavity of the fish.

10. Pull out both the implanter and the plastic straw guide from the incision.

11. Apply to the wound a small amount of oxytetracycline ointment (Terramycin ointment, from Pfizer, Inc., Metro Manila).

12. Let the fish recover in a tank of fresh sea water before placing it back into the cage.
13. Implant fish monthly from February to May.
14. Check monthly for the presence of milt or yolky eggs by following the cannulation technique.

Hormonal Induction of Spawning

Although HCG and LHRHa are two common spawning agents in sea bass, LHRHa is most cost-effective than HCG. Hence the following steps involve only LHRHa to spawn mature sea bass in floating net cages.

1. Spawning cages are prepared, that is, lined with fine mesh hapa net (0.6-0.8 mm or "skin" cloth). Alternatively, spawning tanks may be used.
2. Take out sea bass from its stocking cage or tank and anaesthetize it.
3. Weigh fish.
4. Check the initial egg diameter of ripe females. Also check the presence of freely flowing milt among mature males.
5. There are two effective methods of introducing LHRHa to mature fish; injection or pelleted hormone implantation.

Injection

1. Prepare a fresh solution of LHRHa.
2. An LHRHa dose of 20-100 micro gram of the hormone per kg body weight is recommended. An injection volume of 0.1-0.5 ml per kg body weight may be followed.
3. Draw out enough hormone solution with a 1-ml capacity tuberculin syringe.

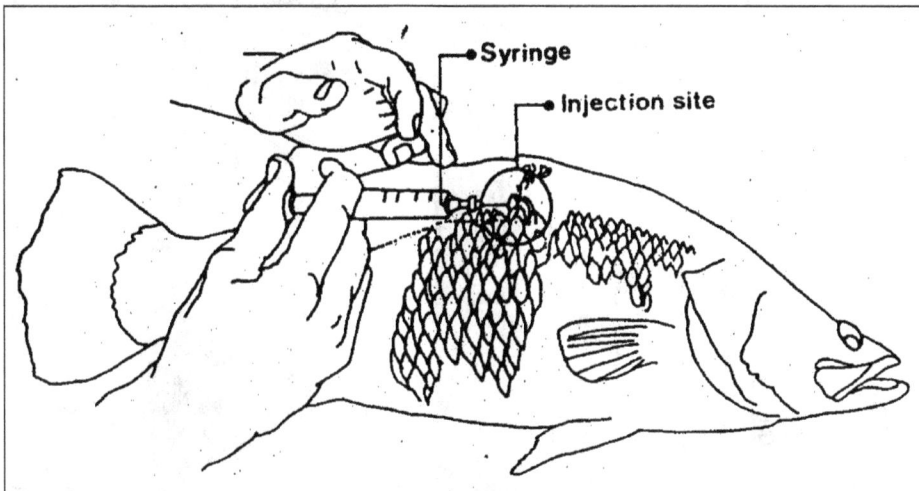

Figure 5.10: Injection of Hormone into the Fish Muscle.

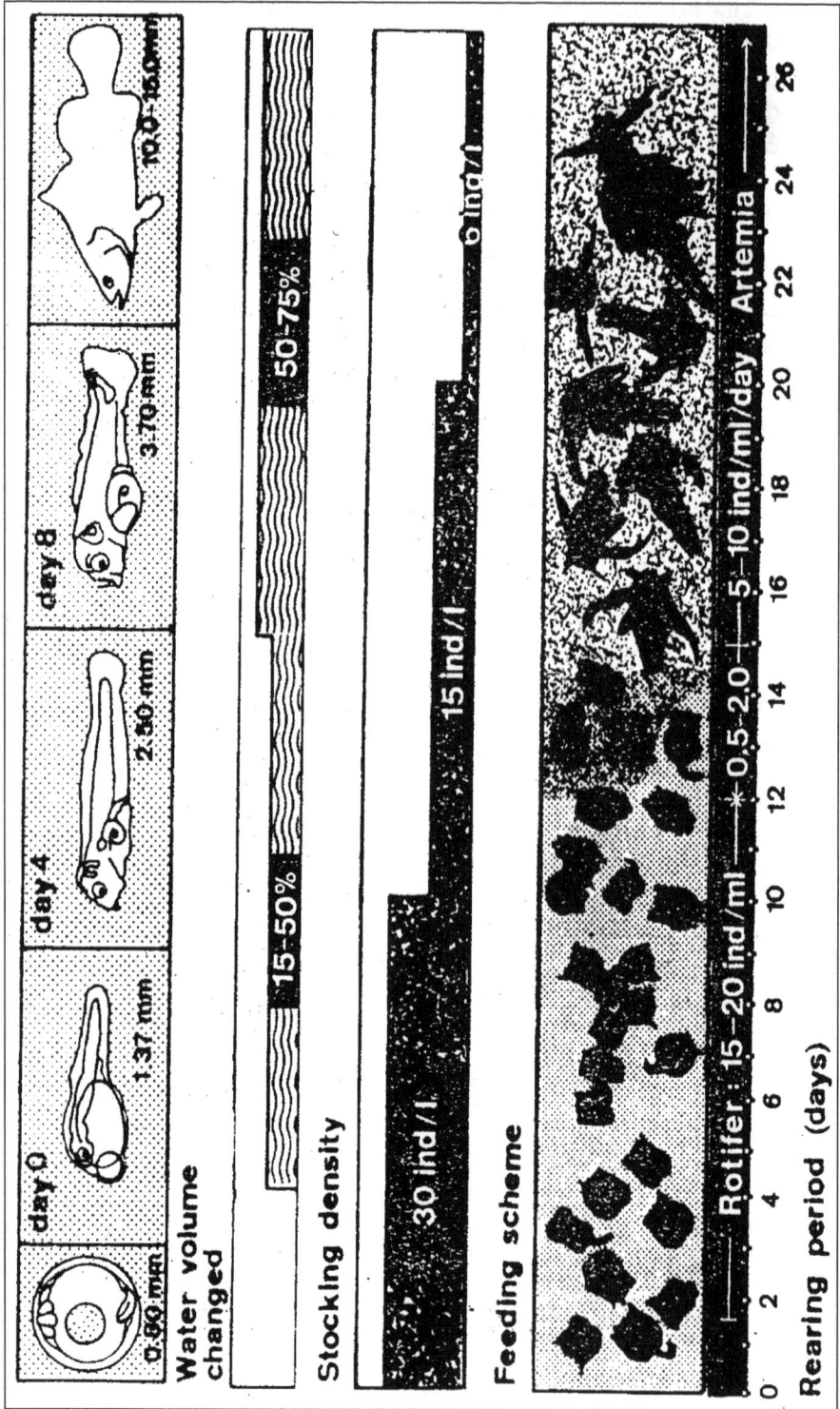

Figure 5.11: Schematic Presentation of Sea Bass Larval Rearing Based on Age of Larvae.

4. Carefully lift a scale with a hypodermic needle of the syringe and inject the hormone at a point 5-10 scales below the dorsal fin of the anaesthetized fish. Prevent unnecessary spillage during injection by gently pressing on the point of injection as the hypodermic needle is withdrawn. Mature male fish may rceive at least 40 micro gram hormone per kg body weight. Inject hormone during daytime.

5. After injection, let fish recover before returning it to the spawning cage or tank. Maintain a 1:2 (female : male) sex ratio in the spawning cage.

6. Wait for fish to spawn two nights after the hormone injection. Mature fish injected with higher than 20 micro-gram LHRHa per kg will spawn thrice consecutively.

Pelleted Hormone Implantation

1. Prepare for implantation; weigh, anaesthetize, measure egg-diameter.

2. Prepare hormone pellets of dose 5-75 micro-gram of the hormone per kg body weight.

3. Implant pellets to ripe male and female fish in daytime.

4. Stock implanted fish in the spawning cage or tank at a 1:2 (female : male) sex ratio. Fish will spawn two nights after implantation of the pelleted hormone. Mature female fish implanted with 5-10 micro-gram LHRHa per kg will spawn once; 20-40 micro-gram per kg, twice or thrice consecutively; 40-70 micro-gram per kg up to four successive times.

Hormone Preparation

Materials

LHRH analogue (luteinizing hormone-releasing hormone analogue)

!7 alpha-methylestosterone (MT)

Clolesterol powder

80 per cent Ethanol	Pellet maker
Weighing scale (accuracy of 0.01 mg)	Pipette (0.1 ml graduation)
Petri dish	Distilled water
Metal spatula	Sodium chloride reagent
Hormone pellets	

Procedure

1. Pelleted LHRHa and MT

(a) Calculate the total amount of LHRHa and MT required when a dose of 0.1 mg of each hormone per kg body weight is used.

(b) Initially dissolve 1 mg MT with 0.2 ml of 80 per cent ethanol. Set aside.

(c) Break open enough of LHRHa ampules and empty the content in a clean dry petri dish.

(d) Combine dissolved MT with LHRHa in the petri dish.

(e) Add 0.5 ml of 80 per cent ethanol to the hormone mixture. Mix well with a metal spatula.

(f) Weigh cholesterol powder following a ratio of 0.2 grams of cholesterol per milligram of hormone.

(g) Combine cholesterol powder with dissolved hormones in petri dish.

(h) Mix well with a metal spatula until the powdered mixture is completely dry.

(i) Calculate and weigh the total amount of hormone contained in the powdered mixture for each fish to be implanted.

(j) Compact the weighed, powdered mixture into cylindrical pellets (3 mm diameter, 3.5 mm height).

(k) Store hormone pellet in separate vials for each fish.

(l) Each milligram of the powdered mixture contains 0.005 mg of each hormone.

2. Pelleted LHRHa

(a) Determine the dose of LHRHa to be implanted to fish.

(b) Calculate the total amount of LHRHa required based on the dose level determined.

(c) Break open the ampules containing LHRHa and empty the content in a clean, dry petri dish.

(d) Dissolve each milligram of the hormone in 0.2 ml of 80 per cent ethanol

(e) Weigh out cholesterol powder following a ratio of 0.2 grams of cholesterol per milligram of hormone.

(f) Combine the cholesterol powder and dissolved hormone in the petri dish and mix well with a clean metal spatula until completely dry.

(g) Calculate the total amount of hormone required for each fish.

(h) Weigh the total amount of powdered mixture containing the hormone required for each fish.

(i) Compact the weighed powdered mixture into cylindrical pellets. Minimize unnecessary loss during preparation.

(j) Store hormone pellets in individual small, capped vials.

(k) Each mg of powdered mixture contains 0.005 mg of LHRHa.

3. Hormone Solution

Procedure

(a) Determine the LHRHa dose level and the injection volume per kg weight of fish.

(b) Calculate the concentration of the hormone solution required based on the dose level and the injection volume.

(c) Calculate the total amount of LHRHa required based on the hormone dose level determined.

(d) Break open enough ampules containing 0.2 mg of LHRHa.

(e) Prepare a 0.9 per cent salt solution by dissolving 0.9 gram of sodium chloride in 100 ml of distilled water.

(f) Measure with a pipette the volume of 0.9 per cent salt solution needed to dissolve LHRHa in order to arrive at the desired concentration of the hormone solution. Swirl the ampules gently to completely dissolve the hormone.

Egg Collection, Transport and Hatching

Egg Collection

Eggs are collected early in the morning (5 to 7 AM). The procedure in collecting spawned eggs from tanks or cages is as follows :

Tanks

1. Check the presence of spawned eggs in the tank by examining water samples collected with a glass container. Fertilized sea bass eggs float and are normally transparent.

2. If eggs are present, securely place at the outlet canal a wooden box with a fine mesh screen bottom.

3. Slowly lower the water level in the tank and transfer the parent fishes to another tank.

4. While the tank is gradually being drained, gently scoop out eggs from the wooden box and transfer them to a 15 liter pail containing aerated fresh sea water. This procedure minimizes impact related stress and mortalities during draining and collection of eggs.

5. Screen out any debris mixed with the collected eggs.

6. Rapidly but gently transfer eggs to a fiber glass tank containing 250 liters of aerated fresh seawater.

7. Refill spawning tanks with seawater and return spawners.

Floating Net Cages

To facilitate egg collection, cages are lined with a fine mesh hapa net installed a few days before the anticipated day of spawning.

1. Check the presence of spawned eggs in the cage by examining water samples in a glass container.

2. If eggs are present, transfer spawners to a spare cage or a large fiber glass tank.

3. Slowly lift the fine mesh hapa net of the stocking net cage. At the same time, splash the sides of the net to wash out adhering eggs. Continue lifting and splashing the hapa net until eggs are fully concentrated to one side.

4. Gently scoop out eggs from the hapa net and transfer to a pail of seawater.

5. Rinse the hapa net cage before installing it back to the stocking cage. Transfer spawners back to this cage.

6. Remove any extraneous debris mixed with collected eggs.

7. Rapidly but gently transfer collected eggs to a fiber glass tank containing 250 liters of aerated seawater.

Egg Transport

If hatchery facilities are located some distance away from the spawning tanks and cages, these steps are followed during egg transport done early in the morning (5-7 AM).

1 Vigorously swirl seawater in the fiber glass tank containing collected eggs. Let dead eggs (white and opaque) settle down; good ones float.

2 Carefully siphon out dead eggs from the tank bottom.

3 Scoop out good eggs with a fine mesh drain net and transfer to a graduated glass beaker.

4 Quickly transfer (100 ml of eggs) to a double lined plastic bag containing 15 liters of fresh aerated seawater.

5 Bubble, then inflate with oxygen until the air space occupies two-third of the total volume of the plastic bag.

6 The plastic bag with several rubber bands and place into a pandan bag.

7 Keep the loaded bags in a cool, dry place. Never expose eggs to heat in such areas like near running motor (in boats or land vehicles) or in open decks receiving direct sun light.

Egg Incubation and Hatching

Incubate fertilized eggs (diameter = 0.80 mm) in 500 liter fiber glass tank at a density of 1200 eggs per liter or less. Provide gentle aeration to keep the eggs suspended in the water column. Hatching will occur approximately 14 hours after fertilization at 28 degree Celsius and 32-33 ppt.

Larval Rearing

Upon hatching, transfer larvae from the incubation to the rearing tank. Estimate the concentration of larvae in the incubation tank and compute for the required water volume of the rearing tank for the desired stocking density.

Stock larvae at an initial density of 30 individuals per liter. Higher densities, namely, more than 90 individuals per liter, may be adopted based on production targets (in terms of fry size and number of individuals harvested) and level of capitalization. Reduce the density to 15 individuals per liter when larvae reach 10 days old and further to 6 individuals per liter when 20 days old.

Types of Feed and Feeding Management

Sea bass begin to feed 50 hours after hatching at 28 degree Celsius. However, it is advisable to introduce the larval feed at an earlier time. If larvae are unable to feed 60 hours after hatching, irreversible starvation will occur and at least 50 per cent of them will die.

Larvae should be weaned gradually to each new food type. This is done by increasing daily the proportion of the new food type while gradually reducing that of the preceeding food. This is necessary so as to train fish to recognize and accept the new feed particle. If weaning is done properly, feed wastage and fish mortality due to starvation are minimized.

Live Food

Larval rearing of sea bass is largely dependent on the use of live food organisms, namely rotifer (*Brachionus plicatilis*), *Artemia* newly hatched nauplii, enriched nauplii, subadult or adult biomass), and the freshwater cladoceran *Moina*.

Approximately 36 hours after hatching of eggs, add rotifers to the rearing tanks at 15-20 individuals per milliliter until larvae reach 12th day. Maintain this density by daily addition of rotifers. Also add 60 liters of *Chorella* in peak bloom to attain a density of 1-3 x 100000 cells per milliliter. *Chorella* is added to maintain water quality and serve as food for rotifers. From day 12 to 15, gradually reduce rotifer density until it is totally removed by day 15.

Wean larvae to newly hatched *Artemia* nauplii over a three day period (day 12 to 14), although weaning may be done as early as day 8 or later depending on food availability and production target. During weaning, provide *Artemia* at increasing densities of 0.5 to 2.0 individual per milliliter per day.

Feed larval fish with enriched *Artemia* nauplii from day 18 to 23 at a density of 5-10 individual per milliliter per day. This is important because sea bass exhibits higher survival rate during metamorphosis when fed enriched *Artemia*.

As the larvae grow bigger, they ingest larger feed particles. They are fed with subadult or adult *Artemia* biomass at one individual per milliliter or higher.

As an alternative to *Artemia*, *Moina*, a freshwater cladoceran, may also be fed to 25-day old sea bass at not less than one individual per milliliter density. When using *Moina* as feed, lower the salinity of rearing water to approximately 10 ppt and feed at least 4 times a day.

Non-live Feeds

Some non-live feeds used include frozen *Artemia* biomass, trash fish, and artificial diets. These feeds can, in excess, cause rapid deterioration in the quality of rearing water. Thus, food ration and feeding frequency must be controlled to avoid water fouling.

Prepare frozen *Artemia* biomass by freezing freshly harvested sub-adult or adult *Artemia* in plastic bags, each to contain the amount needed for one day feeding. During feeding, break the frozen biomass into small pieces and distribute in the tank. As it thaws, the *Artemia* particles will slowly separate allowing the sea bass larvae to swallow each piece whole.

Only fresh trash fish should be used. Remove head, entrails, and bones, then chop to fine bits the remaining flesh. During feeding, give trash fish slowly to allow fish ample time to feed before the trash fish particles sink to the bottom. Feed at least 3 times a day.

Artificial diets hold promise for sea bass fry production However, the feeding management has not yet been standardized and the economics is still unknown.

Water Management

Starting on the 4th until the 14th day of culture, drain and replace daily 15-50 per cent of the water volume. On the 15th day, when feeding level with newly hatched *Artemia* nauplii is maximum, change daily 50-75 per cent of water. Once feeding with non-live feed begins, change 100 per cent of water daily.

During water exchange, drain water in the rearing tank using a siphon with its inlet covered with plankton net. This is to prevent larvae from being drained. Refill tanks to its original volume with new sea water. In addition, siphon off feaces and other debris found at the tank bottom every morning from the third day until harvest.

Size-Grading of Larvae

Cannibalism among sea bass reduces fry survival. Hence grade stocks often to remove "shooters" and reduce the chances for cannibalism. Usually, cannibalistic fish swallow their prey whole. Since the maximum size of prey that a cannibal may ingest is approximately 60-67 per cent or two-third of its length, separate fish with length difference of 33 per cent or more.

During size grading, use a wide hose (inner diameter = 5 cm) to siphon of fish into sorter boxes. Arrange the boxes serially so that the net with the biggest mesh size is the innermost, while the finest mesh net is outermost. When siphoning is completed, gently lift and lower the innermost box repeatedly until fish that remain inside are too big to pass through the net. scoop out these 'shooters" and rear them separately. Do the same for the next sorter box. Fish that remain in the second box should constitute the average-sized population. Fish that pass through the first two boxes are retained in the outermost box and are smallest individuals.

The nets may be changed accordingly to obtain different mesh sizes. Siphoning is possible only if the rearing tank is at a higher elevation than the sorter boxes. If this is not the case, simply drain water in the rearing tank and use a scoop net to gather and transfer fish to the sorter boxes.

Usually, shooters begin to develop several days after *Artemia* feeding is started. Therefore, the first grading should be done at this time. Grade subsequently whenever shooters are evident.

Harvest and Transport of Fry

Twenty-one-day old sea bass fry, approximately 1 cm total length, can be harvested from the hatchery for rearing in nursery systems. Transporting them to the farm site is easy since the fry is quite hardy.

Harvest the fry by partially draining the water in the tank. Using a fine-mesh scoop net or small basin, scoop the fry and transfer them to containers, preferably big white basins. To estimate the number of harvested fry, put a known number of fry in a basin with a known volume of water. Use this model for visual estimation of the number of fry in similar basins. Distribute the fry into the other basins containing same volume of water as the model and visually compare the number of fry in these basins with the model.

Pack the fish in double lined plastic bags filled with fresh seawater. Inflate the bags with oxygen at water to air ratio of 1:2, seal with rubber bands, then put the bags inside buri bags or Styrofoam boxes. The density normally used during transport of 1-cm fry is 500 fry per liter or 5000 fry per 10 liters of transportwater. However, as much as, 1600 of 1-cm fry per liter can be packed for 8 hours of transport at 28 degree Celsius. For practical purposes, always adjust he loading density based on the duration of transport and size of fry to be transported. Do not feed fry at least 24 hours before transport so as to reduce production of metabolic wastes. It is best to transport fry during cooler periods of the day, that is, early morning or late in the afternoon.

Production of Larval Food

The larval food types utilized for sea bass fry production include green algae (*Chlorella, Tetraselmis*), rotifer (*Brachionus**, brine shrimp (*Artemia*) and *Moina*. Mass production of green algae and rotifer is normally done in big canvas tanks, whereas *Artemia* and *Moina* production is normally done in 250 liter hatching vessels and 1-2 ton fiberglass tanks.

Chapter 6

Breeding in Fish Culture

In fish culture, breeding of fish is carried out with the primary object of modifying their hereditary properties in the direction required. The process of reproduction in fishes to a large extent controlled by man. Such control is possible only to a limited extent in natural water basins, but in pond fish culture it is possible and even necessary to utilize all modern methods of domestic animal improvement with due regard to the peculiarities in biology and breeding methods of the fishes in question. Breeding practiced in pond fish culture also ensures the provision of spawners necessary for obtaining commercial products of good quality in required quantities.

Although very important in their role, selection and breeding are at a rather low level of development in pond fish culture as compared to other branches of animal breeding. Even among carp, which are the oldest cultivated fish, only few species and breed groups compare with the highly specialized breeds of domestic animals. This is to a large extent due to the fact that carp is not a completely domestic animal. Also the methods of breeding and management have been far from perfect. The considerable rate of genetic plasticity if carp is of great value, as is the wide range of its paratypic variability.

In fish culture, there is certain advantages as compared to other types of animal breeding. Fishes have high fecundity, which allows the use of strict norms of selection. Fertilization is external in the majority of fishes and this enables exercising of control over their embryogenesis.

Selection-Breeding Farm

Selection-breeding farms are highly specialized farms. Besides production of new breeds, breed groups and improvement of the existing breeds, these farms would conduct research on selection methods on the system of pedigree and commercial cultivation (paying special attention to the problems of attainment and fixing of heterosis) on methods of spawner evaluation and estimation by generations and by the utilization of hybridization. These farms would develop scientifically sound biotechnical methods of raising and maintaining pedigree fish. These problems require long term and close studies and repeated experiments, for obtaining reliable results. The selection-breeding farms should, therefore, be established as highly specialized farms, the staff of which should consists of specialists in genetics and selection, and work under the guidance of competent scientific bodies. Their pond facilities should consist not only of hatchery ponds for rearing spawners, holding and quarantine ponds, but also a system of numerous experimental ponds. When designing such farms it is necessary to provide for a hatchery in order to obtain fish eggs and fry as well as equipped laboratory, aquarium rooms etc.

In the structure and capability of selection breeding farms several gradations are possible depending on their specialization towards selection (for example, resistance to diseases in relation to local conditions and requirement etc). It is clear that creation of standard projects for such farm is not feasible, but special methodological directions have been worked out in USSR by the Scientific Research Institute on lake and river fisheries.

Parent Fish Production Farm

One of the main considerations of raising stock of parent fish is the reproduction of two non-related breed groups or strains for producing parent fishes with selected characteristics of males and females of different origin. Such a method of commercial breeding has been suggested by Shaskolsky (1954). It excludes the possibility of inbreeding (which is not permitted in commercial fish culture) and allows manifestation of the effect of heterosis. Commercial hybridization of selected carp with Amur wild carp and crossing of two strains of Ropsha carp yielded interesting instances of high heterosis (Kirpichnikov, 1966).

The tanks of breeding farms are not limited to production of pedigree fish. It is necessary for these farms to carry out continuous work on further improvement of the breed-groups received, mainly by means of mass selection. Mass selection is one of the most important methods of fish selection. The selection should be strict and directed. Carp selection is carried out in the breeding farms in three stages among the yearlings, at the age of harvest for commercial use; and at the time of transference from the reserve to the spawning stock at maturity. The ratio of selection intensity (number of individuals left for breeding related to the number reared) should not exceed 50 percent for the yearlings, 10 percent for two-year old fish, 25 percent for young females and 50 percent for young males. Thus, the general ratio of selection intensity for younger groups in carp nurseries should not exceed 5 percent as recommended by Schaperclaus (1961). Introducing additional selection among spawners, bringing the selection intensity up to 1.25 percent for females and up to 2.5 percent for males.

At the first two stages (yearlings and two-summer old) the weight of the fish is the criterion for selection, though among the two summer old, exterior features are taken into consideration. At the third (parent fishes) the degree of expression of sexual characters is considered in the selection.

This simple method permits oe to estimate the quantity of replacement the parent fish required for obtaining the number of spawners needed, and to calculate the necessary number and area of ponds for separate storage and rearing of replacement of parent fishes of all ages.

When designing breeding hatcheries one should not limit the farm plan to the production of parent fishes only. This is not profitable and, besides, the availability of commercial ponds is useful for checking the quality of pedigreed fish being released. Thus a breeding nursery is a kind of specialized pedigree department of a commercial farm, where release of parent fishes is planned along with commercial production. In the breeding farms as well as in the selection-breeding farms of the highest type there should be provision for quarantine ponds and a hatchery. The staff of breeding farm should have a specialist on fish selection and their work should be in close connection with selection-breeding farms of the highest type in the system in which they are included.

When arranging pedigree activities in carp breeding, the generally adopted principle of correlating the work on modification and improvement of genetic structure of stocks with the creation of suitable conditions for raising and keeping is followed. This is absolutely necessary in order to maintain adequate physiological condition of fishes and to obtain the required profitable production. Therefore, certain fish cultural standards have been introduced for keeping the parent fishes. These standards provide for optimum densities of stocking for ensuring the necessary profit, conditions of feeding fish and pond fertilization for providing a certain amount of natural food in the ration.

Two year cycle is followed in carp farming. High stocking rates are used for raising young-of-the-year and two-summer fish, which permits with proper fish cultural techniques, the growth of fish of required commercial standard weight, and at the same time ensures high production in the ponds. For raising young-of-the-year and two-summer fish for replacement of parent fishes, such stocking densities are not allowed; moreover it is considered incorrect to provide these fish with extraordinary conditions of life that differ greatly from commercial conditions. Therefore, for the young-of-the-year and for two-summer fish, rather moderate stocking rates are adopted. They are intended for production of increased gain as compared to the commercial norms, and not for the attainment of maximum possible gain.

For the reserve stock of all ages, separate ponds should be provided. This is necessary not only for better control over the growth and condition of fishes, but also for adequate feeding. When various age groups are mixed, certain confusion is inevitable and older fish of low quality may be confused with the better and younger fish. For two breed-groups raised in a breeding farm, separate ponds are provided, but it is possible to alternate their usage each year.

For production and replenishment stock pair of parent fishes (a pair = 1 female and 2 males, the usual ratio for natural spawning in carp breeding), the number of young-of-the-year being raised (and accordingly the area of the ponds) is not regulated, but it is necessary to rear not less than 25000 young-of-the-year. This number being higher as the strictness of selection becomes more severe. Therefore, it is not necessary to design special ponds for the young-of-the-year if in the commercial part of the farm there are proper sized ponds available for rearing and wintering, out of which it is possible to choose the ponds necessary for rearing breed generations from the selected parent fishes.

Reproduction of the pedigree stock in a breeding hatchery should be planned keeping in mind the possibility of an annual change of 25 percent of the spawning stock.

Commercial Farms

These are farms with stocks of parent fishes for reproduction, but which do not raise their own replenishment stock. They receive the necessary parent fishes from the breeding farms. The purely commercial fish farms do not require experts in selection, but the fish farmers should be qualified enough to tackle the main problems of breeding, to eliminate properly the quality of parent fishes when judging and matching them for spawning as well as to make adequate use of bio-technical regulations in fish rearing.

For purely commercial farms, ponds for replenishment stock are not necessary, but there should be holding ponds (summer and winter) in which the parent fishes have normal conditions for fattening and ponds for fish to replace loss of parent fishes.

The quantity of commercial fish (two year old fish) by weight obtained in one season as the result of spawning, one carp female (or pair) is conventionally taken as an indication of its productivity. This value is summed up from the active fecundity of a female (number of fry transferred for rearing); survival in the process of rearing and adopted standard piece weight of commercial fishes. For example, by the standards adopted when estimating the capacity of a pond farm design, the following figures of female productivity is taken into account.

1. Active fecundity – 125000 larvae;
2. Young-of-the-year yield – (75 per cent) = 94000;
3. Yearlings yield – (85 per cent) = 80000;
4. Two-year-old yield – (90 per cent) = 72000

Thus at a standard weight of 0.5 kg for two-year-old fish the annual commercial production of a female is equal to 36000 kg.

Active fecundity of carp females can reach the level of 400000 active fry, not only due to the fecundity of the female herself, but due to the efficient incubation methods and fingerling rearing methods used, periods of transferring etc. However, the importance of the female's fecundity *i.e.,* evident and practical work carried out has proved that good females can give an annual yield up to 100000 kg or even higher.

High productivity of females permits one to operate with a few parent fishes, which in turn, facilitates the establishment of better conditions and permits control over the stock and selection of the best individuals. Surplus parent fishes do more harm than good for production. The spawning stock (and consequently, the replenishment stock) should be kept only in the quantity desired with a certain reserve, the amount of which is determined in relation to the health of the individuals. In a healthy farm it is sufficient to have a reserve of 25 to 50 percent. Such control over the stock of parent fishes and of the reserve is essential in purely commercial farms and in breeding nurseries for efficient management of work.

Inventory of Fish for Pedigree Fish Culture

Method of Inventory

One of the important measures of regular control over the condition of pedigree fishes and brood-stock structure consists of adequate arrangement of annual inventory of the whole stock of parent fishes in purely commercial farms and of parent fishes and reserves in the selection breeding farms. Properly arranged inventory takes into account all modifications in the brood stock indexes, which appear as a result of selection work and under the influence of the environment as well as those essential for carrying out selection.

The accuracy of their determination effects the correct estimation of the pedigree fish's value as well as the formulation of adequate methods of improving them.

Inventory is carried out in spring in the USSR, directly at unloading of the wintering ponds. Proven facilities should be prepared before hand; that is, scales, measuring board, intermediate reservoirs for fish, sorting tables etc. The fish should be handled with extreme care, female parent fishes in particular, avoiding any kind of injury. For example, all the measurements should be taken with the fish in a secure and stable position. For this purpose, measuring boards with two vertical sides are especially convenient. For weighing a special container is necessary; a kind of cradle for one fish at a time is used in keeping with the size of parent fishes.

The whole group of parent fishes undergoes individual visual examination, during which all sick fish are rejected, as well as the fish with malformation or serious injuries.

The inventory is started on the parent fishes, separately by breed-groups and sex. Individuals are sexed and sorted into classes. In the most simple cases, only the females are separated (they are placed in separate ponds) into three classes;

☆ I class – The best individuals in respect of all characteristics, from which offsprings are to be produced;

☆ II class – Fish with less desirable characteristics than the above; reserve for the second turn utilization;;

☆ III class – Fishes to be substituted for the reserve, if and when necessary, they are removed from the stock on the termination of spawning.

Individual indexes, determined and registered during the inventory (and put down into special logs, kept in the farm) include the following important data; breed group, sex, age, class, fish tag (group, individual). Colouring peculiarities or distinctive marks, in the carps, the type of scale, the degree of expression of sex characters and readiness for spawning, and individual weight and measurements data by which the indexes of exterior features are determined. All indications which are the cause of immediate rejection are registered.

Special attention is paid on the following indications.

Disease and Malformation

It is desirable at big farms and breeding farms, that an expert on ichthyo-pathology should participate in the processing of inventory, the opinion of this expert being decisive if there appear any doubts concerning rejection of diseased fish. Fish with malformation (in the fins, head, body proportion etc) are excluded from the pedigree fund intil they reach maturity. Malformed parent fishes should be rejected. Schaperclaus (1955) was right to stress that all fishes with inheritable defects should be excluded from the stock. In particular this relates to the malformation of the bladder, which sometimes result in lopsided fishes, as seen frequently among adult carp.

Sex and Maturity

The first class should embrace only such parent fishes whose characters are obvious. This is specially important when transferring young fish from the reserve. The majority of carp rejected in farms during rather severe selection of youn fish, are young females which are slow at maturing. One must take into account the difference in significance of late and early maturity in different climatic zones, because in hot climates late maturity is a positive feature.

Age Groups

The most productive age group of carp in moderate climates is considered to be 6 to 11 years for females and 5 to 10 years for males. It is not recommended to refer young females, just transferred from reserve to the first class, but they should be kept in reserve and permitted to spawn even when their participation in the spawning appears unnecessary.

Weight

Weight is to be adopted as the main indicator in mass selection. But the selection of best parent fishes for the first class is carried out among fish of various ages, thus the weight is not a decisive factor. For purpose of estimation, the expected productivity of females and their gain in weight in the past season are of importance; for example, a big female which has lost weight in comparison to its weight the previous year has no advantage over a female of smaller weight which has shown marked gain in weight.

Loss in weight as well as poor growth in the absence of other symptoms, usually indicate that the fattening conditions for the females in the post-spawning period were not satisfactory; due to malnutrition. The females fail to make up the spawning losses and are not able to accumulate reserve materials for the next spawning. Therefore, it is recommended that carp spawners be fed with fodder mixture in the pre-spawning period also and to refrain from too dense stocking of parent fishes after spawning for summer fattening. With adequate feeding and separate rearing it can be accounted for normal gain (1 kg or more during a summer period, when stocking 100 to 150 females and 150 to 300 males per hectare.).

External Characters

Much attention has been paid to the external characters in carp breeding, and at one time they were considered as the foundation for the establishment of race (breed) standards. Many scientific investigations were carried out to find direct connections between the external characters and rate of growth. They were taken into account when selecting the reserve and the parent fishes, because they are related not only to the species or race peculiarities of fish constitution, but serve as the indications of normal development and growth. Three minimum indexes were determined for carps;

1. Relative body height – L/H, where L is standard length and H is the maximum body height measured at the dorsal fin;
2. The maximum body width – Br/L, where L is the length and Br is the maximum body width;
3. Coefficient of condition – k = g/L cube. 100, where g is the weight of fish and L is the length of fish.

Coefficient of Conditions

The index of coefficient of conditions, k is adopted in carp breeding as an indication of expected winter-resistance of the young-of-the-year fishes. This index is useful for the evaluation of other species of pond fish as well. Parent fish selection by the values of k ensures improved stock in respect of other characters too. Using auxiliary tables, it is possible to determine k during inventory under field conditions, proceeding from the data on weight and length, the highest appraisal is given to the parent fishes with the highest k value (within the limits of each species or race group).

When estimating by k, and specially, L/H, it is necessary to ensure that good indexes (high values of k and low of L/H) should not be conditioned by defects in constitution; spine distortion, caudal peduncle shortening etc. These malformations, very often hereditary, are rather characteristic of cultivated carp, probably their occurrence is related considerably to the selection for body height, to western European fashion for round, fit-to-plate carp.

Determination of individual indexes is necessary for young parent fishes being transferred from the reserve to brood-stock ponds. In younger age groups individual indexes are determined by processing farm average (without selection) samples. It is necessary to determine piece weight of all reserve categories (sum weighings are used for this purpose) and, of course, to ensure their accurate registration.

No less important than spring inventory is autumn registration of fattening results. Usually at autumn fishing individual weighing and measuring of fish is not done, unless required for some program of special work It is most desirable to weigh parent fishes before and after spawning. This allows one to estimate approximately the actual fecundity of females and to register precisely the results of summer fattening.

Breeds of Fishes

The breed can be defined as a man-created population which is characterized by certain heritably fixed characters (level of productivity, biological features, external and morphological characters). The main properties of a breed manifest themselves in their most typical form under the conditions which existed when this breed was created. It concerns, particularly, the physiological characters (growth rate, food conversion etc.).

Breeding in Fish Culture

Though fish culture has been in practice for centuries in Europe and thousands of years in China, stable and specialized strains have not been produced as yet even among carps, which are the only group of pond fish that have been subjected to experiments in selection.

Carp culture in China constitute a part of polyculture species. There is a diversity of forms those which differ in colour, external characteristics and peculiarities of scale covering. They are evidently indigenous breeds. Their differentiation is based on heterozygosity of these characters, which are inherent in carp rather than on purposeful selection. Homologous genetic variation controlling scale patterns in European, South Asian and Japanese carps derived from different sub-species of wild carps was discovered (Kirpichnikov, 1967, Djan din Djong, 1967).

Breeds of Carps

According to the pattern of scale covering, four basic forms (races) of carps are distinguishable; completely covered with scales, with scattered scales, with scales arranges in line, and scale-less; which are reffered to scaled, scattered, linear and leather carp.

These four categories are determined by the combination of two pairs of independent hereditary factors (S-s and N-n genes), which are obviously identical for European and Asian carps.

According to a body shape index or conformation factors, that is, the ratio between standard length and maximum height, the west European carp has been attempted to classify and even to standardize. In the case of cultured carp the index was assumed as ranging from 2 to 3, and carp were divided into high-backed (l/h = 2-2.6) and wide backed (l/h = 2.6-3).

After the First World War, the German Agricultural Union (D.L.G) made an attempt to establish standard factors for the four major races which had been recommended for breeding. These indices included almost all the above characteristics of body shape and also the features of scale covering.

The high-backed Aischgrund and Galician carps were rated highest in value (it was suggested that there is a direct relationship between growth rate and body height), followed by wide-backed Lowscatian and Frank carps.

According to the pattern of scale covering, these standard forms were assumed; Leather for Aischgrund and Frank linear for Galician and scaled for Lowscatian.

Along with investigations into genetic variability of carp, attempts were directed at creating the Ukrainian carp (framed and scaled). This work was carried out by Ukrainian selectionists headed by A.I. Kuzama.

Ukrainian carp belong to high-backed breeds. When food supply is adequate the conformation factor reaches 2.2-2.3 (or 2.0 sometimes), and there is a good growth and high production.

Typical farmed carp have the scales arranged in the shape of a frame composed of a dorsal row and patches located near the head, on the tail and near the anal and ventral fins. Their disadvantage is low resistance to infectious dropsy (a serious and widely spread infection), besides their heat loving and do not satisfactorily withstand the winter in northern areas.

Creation of cold resistant carp breeds is one of the important problems of carp culture in the USSR. For this purpose, the most promising race of wild carp, Amur wild carp (*Cyprinus carpio haematopterus*) is the best form for use in hybridization of carp. The hybrids of the first generation have gained recognition as a commercial breed. They are noted for remarkable cold resistance, high viability and fast growth rate.

The heterosis effect in growth and viability of these hybrids can be observed in the first year (especially during the first months) of their life. The principal disadvantage of this commercial hybrid is that both the initial forms (selected carp and wild carp) should be kept at rearing farms because the hybrids are not preserved for further breeding.

Under the conditions of commercial hybridization conducted on a large scale, the stocks become intermingled and naturally the efficiency is lost.

Selected hybrids of cultured carps X Amur wild carp have been produced to the fifth generation (applying special scheme of crossings) and resulted in the creation of a new breed group which gained the name Ropsha (after the name of experimental farm where the work was conducted) or Northern carp. These carp now constitute the main pedigree stock for the north-west region of the Russian Federation and are cultivated in Estonia and Siberia. In the Ukraine, when crossed with local carp they showed a strong heterosis effect.

In the South, Ropsha carp grow well, and they and their hybrids between local strains are more resistant to dropsy than local carp. Ropsha carp have the entire scale covering and genetically are homozygous. Carp with scattered scales participated in initial crossings. In the process of selection, a special segregation check was performed with a view to choosing homozygous producers. In diagnostic characters these carp occupy a place between common carp and Amur wild carp. The conformation factor (l/h) in parents varies within the limits 2.9-3.3, while in Amur wild carp it ranges from 3.2 to 3.6 and in the common carp from 2.7 to 3.0. Ropsha carp are not inferior to Ukrainian carp in respect of fecundity.

Chapter 7

Genetics in Reproductive Biology

Inheritance and Variability

Genetics is the science of inheritance and variability. Inheritance is the capacity of the offspring to acquire the characters and peculiarities of development of the parents. Variability is contrary to inheritance, being the ability to change hereditary factors, as well as the manifestation of their properties in the process of development.

The most remarkable feature of the reproductive process is that the young shows significant similarity to their progenitors, all species of plants and animals producing offspring resembling themselves. Individual features are inherited by each generation with great precision.

Members of the same family will always differ from each other and most or even all of them will differ from their parents,no matter how great resemblance exists between them. In some cases the variability related to distribution of different characters among offsprings operates according to definite laws of inheritance, while in other cases it is related to environmental conditions. Variability due to new gene combination is called combined hereditary variability. In the process of development and vital activity of the organism influenced by both external and internal conditions, hereditary changes appear due to changes of genetic structure and this is mutational variability. When variability is due to a modified manifestation of a character, due to environmental conditions, it is called modification. Here it is not the definite manifestation of the character having a modified variability that is inherited, but the whole range of variation of the character depending on the environment- the so-called reaction norm of the organism.

Strictly speaking, there are no hereditary changes, but all the character changes are hereditarily determined. Inheritance is the phenomenon which not only ensures resemblance, but also variations of organisms in a number of generations. These

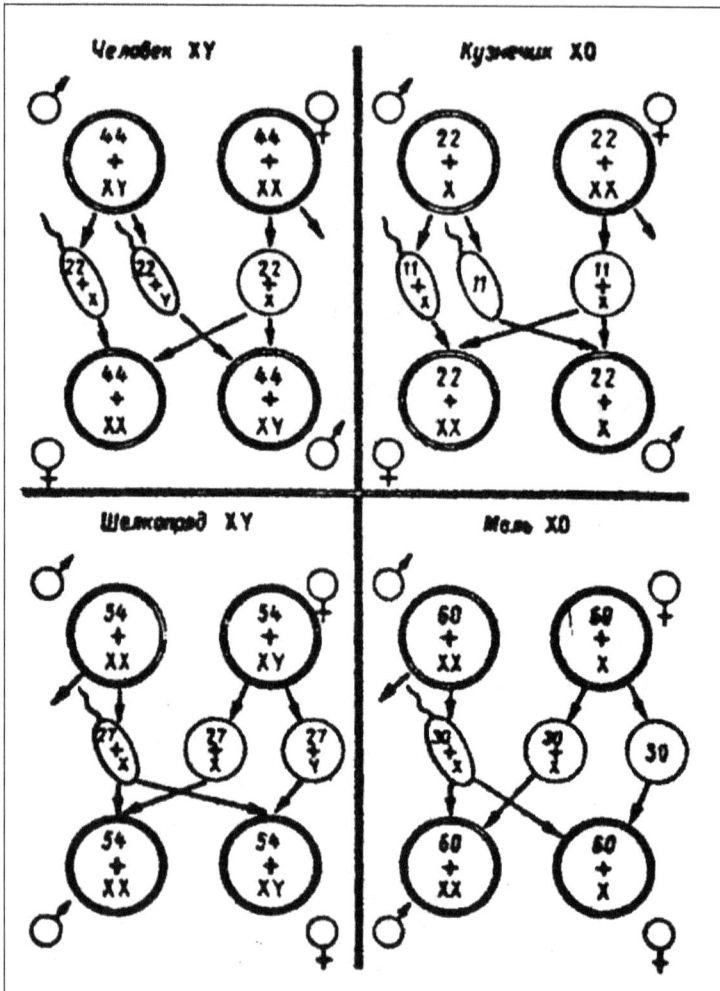

Figure 7.1: Scheme of Different Types of Chromosome Determination of Sex.

variations appear as a result of variability of hereditary characters. Therefore, inheritance and variability are the two parts of the same process which makes for the evolution of organic forms.

Thus, the difference between hereditary variability conditioned by a genotype and maintained in a number of generations, and non-hereditary variability which is modified changes in the phenotype of the organism is understood.

Theory of Inheritance of Acquired Characters

The theory of inheritance of acquired characters, put forth by J.B.Lamarck in 1809 faced criticism and was proved to be wrong by the development of experimental investigations on population genetics.

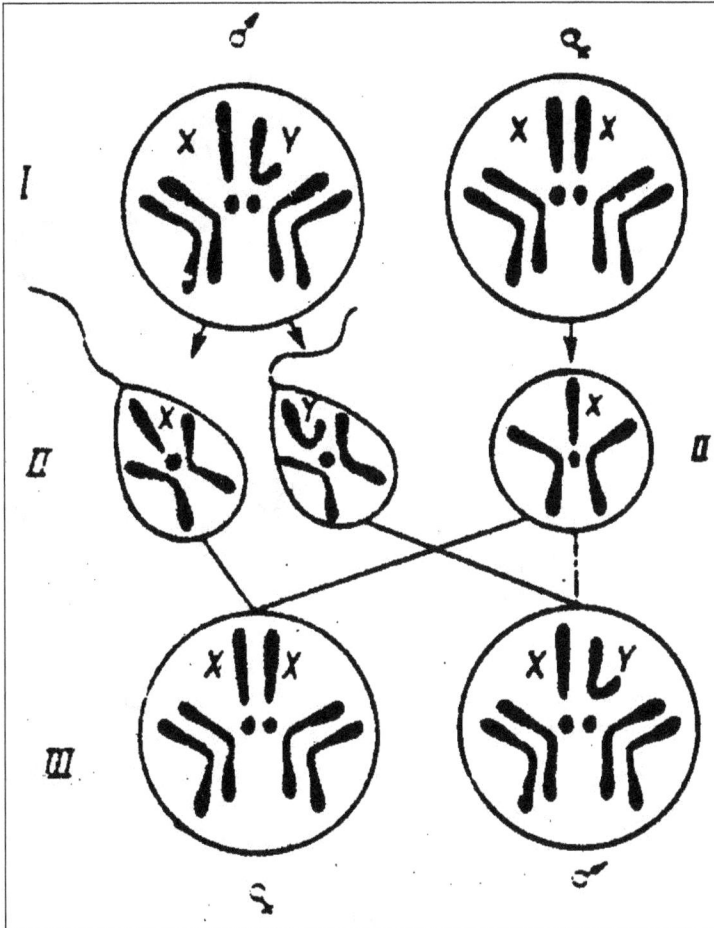

Figure 7.2: Diagram of Sex Determination with Male Hetrogamy (xx-xy) *Drosophila*. I: Diploid tissue; II: Spermatozoal; II': Egg-cell; III: Offsprings

Following the logical sequence of the statements (Weismann (1883-1892) pointed out the differences in the plasma of generative and somatic cells. The generative (germ) cells are the only ones that participate in sexual reproduction determining continuous transfer of hereditary characters from generation to generation. Hereditary characters of an organism are determined by germ plasma, while the hereditary changes are determined by the changes in its molecular structure; hence Weismann theorized on the non-inheritance of characters acquired by the organism during its life cycle.) it might be admitted that the character change under the influence of environment leads to adequate change of hereditary structures (genes) regulating this character. Such a change should have brought about the hereditary fixation of a given character. Though mutations (hereditary changes) result from the influence of environment, they do not appear to be adequate to bring about hereditary fixations. Mutations occur in many directions; they can be adaptive for the organism since they take place in separate cells, their action being manifest in

So, studies in the field of genetics should deal with both inheritance and variability, two opposing and closely inter-connected processes, as well as the material bearers of hereditary information. The investigations should cover all structural levels of living organisms; molecular, chromosomal and cellular, as well as the organism as a whole and its populations.

Genetic Analysis

Genetic analysis is generally based on the study of characters developing in a number of generations.

The principal means of genetic analysis is hybridologic analysis which is applicable at all structural levels of the living matter, from molecule to body. This method is used in the investigation of hereditary structure, behaviour in various systems, interaction of genes and incorporation of characters by genotypes. Hybridologic analysis makes it possible to establish all the principal concepts of genetics.

Cytological methods help to investigate material bases of inheritance, namely, anatomy of hereditary structures.

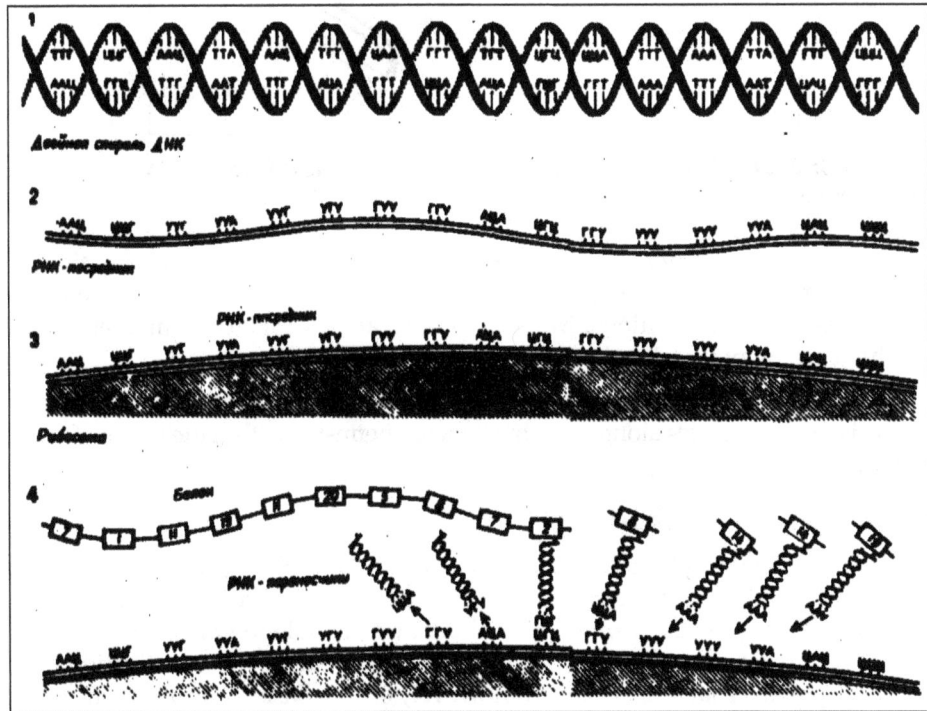

Figure 7.4: Genetic Information Coded, in the DNA Molecule (1) is the basis of protein biosynthesis. It is reproduced in RNA, the messenger (2). SSA as a messenger passes on to ribosomes (3). Amino acids, designated and numbered as rectangles, are transferred to corresponding sites on RNA-messenger by the molecules of RNA-carrier (4). When connected, certain amino acids form a protein molecule.

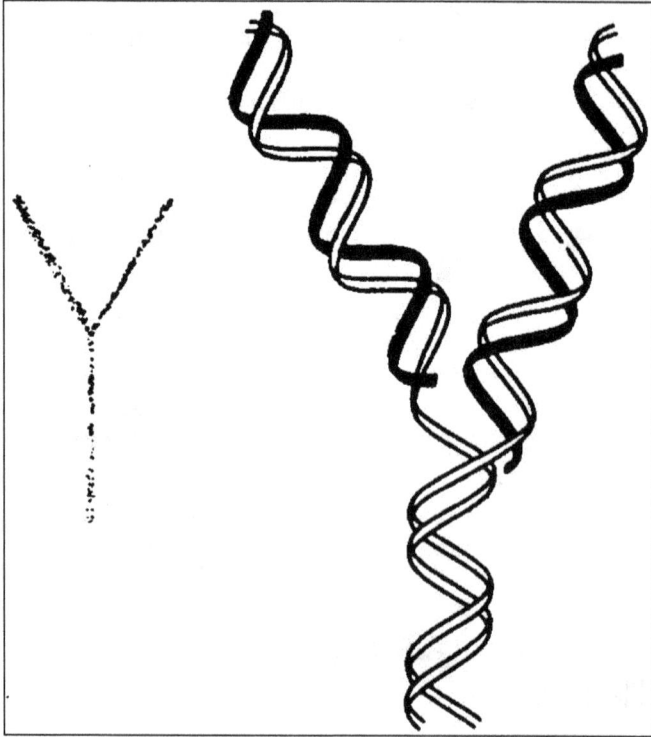

Figure 7.5: Double DNA Molecule; the Formation of a New Thread follows Molecule Segregation.

Cytogenetic methods came into being as a result of the combination of cytological and genetic methods. As such, it analysis the behaviour of hereditary structures in cell generations. Modern genetic analysis includes mathematical, biochemical, biophysical and other methods of investigation.

The possibility of application of exact qualitative and quantitative methods of analysis places genetics alongside physics and chemistry within the range of exact natural sciences.

Development of Genetics

Regularities of inheritance were first detected in plants, and that was not incidental. Plants are easier to study than animals for inheritance regularities. It was only during the 20th century that it became possible to study and pinpoint the characteristics in animals that were much more suitable for investigation than in many plants. That was achieved, however, on the basis of application of the discoveries made on plants.Genetics as a science came into being as a result of studies on animal and plant breeding.

By selecting and crossing the best representatives of any group from generation to generation, relative groups, that is, strains, breeds and brands having their own characteristic hereditary traits, were created. Originally observations and

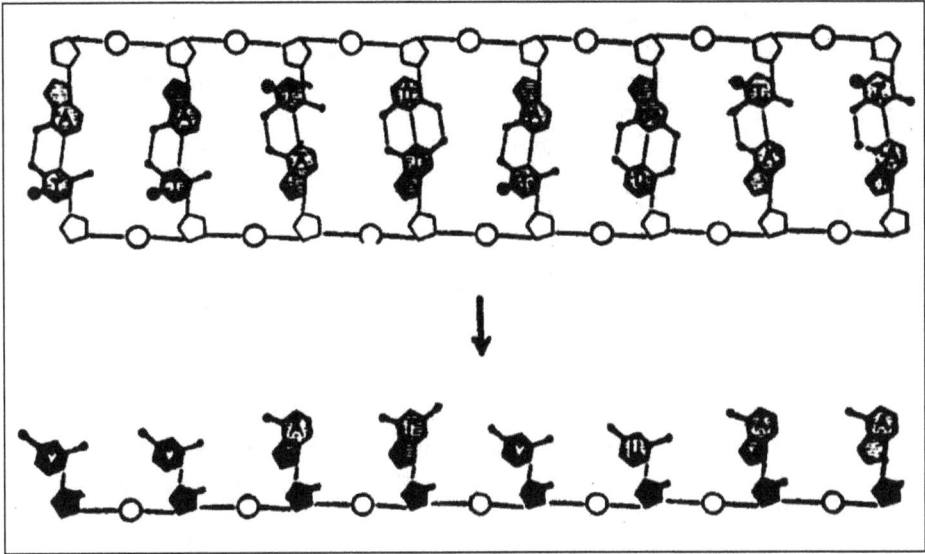

Figure 7.6: DNA Molecule.

comparisons were too few to be formulated into an independent science. In the second half of the 19th century, however, the rapid development of cattle breeding and plant breeding (seed-farming in particular) resulted in increased interest in problems of inheritance.

The first systematic work on inheritance and variability problems was Darwin's "Origin of species". The author managed to collect a large amount of data, though Darwin failed to establish regularities of inheritance at that time.

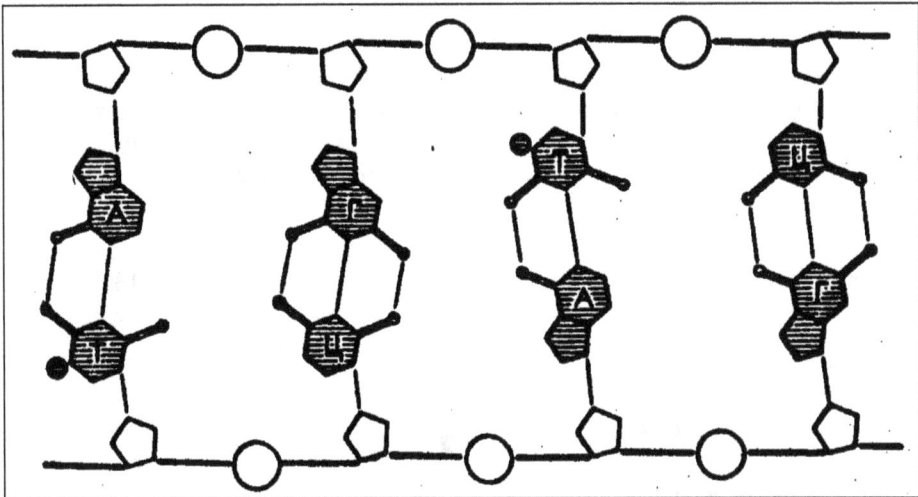

Figure 7.7: DNA Molecule Resembles a Rope Ladder Coiled into a Spiral, where Pairs of the Bases Connect Two Parallel Chains Consisting of Deoxyribose and Phosphate Units. In all pairs, adenine (A) is connected with thymine (T); and guanine (G) with cytosine (C).

The progress of experimental biology at the end of 19th century was of great importance in establishing genetics as science. In the field of cytology basic regularities in processes of mitosis and meiosis were discovered, as well as the permanent number of chromosomes in cell nuclei of each species. T. Boveri, E. Van-Beneden, *et al.*, proved that gamete fusion at fertilization restored the original number of chromosomes which are constant for each species. They proved that in a zygote, the first cell of future organism, half the number of each parent's chromosomes is incorporated.

At the end of the 19th century, a few theories were formulated which attempted to explain the phenomenon of inheritance. All of them are only of historical interest. Weisman's theory of "germ plasma and its continuity" (one of these theories), he put forward an explanation for reduction division, the process of which was not clear then. Maintaining his theory of natural selection in 1886, Weismann postulated the role of sexual reproduction in hereditary variability. He suggested that the latter was conditioned by a mixture of different germ plasmas which occurs at sexual reproduction. It is this very process that provides for the constant occurrence of individual hereditary changes in offspring, which is the material for natural selection. Weismann suggested an explanation of the essence of two reduction divisions preceding each conception. The process of reducing the number of germs inherited from ancestors is considered by Weismann to be the reason for constant change of germ unit sets that get into the organism in any generation. Thus, it is these processes of reduction division with the subsequent fusion of each parent nucleus occurring at every sexual reproduction that is responsible for endless diversity of inherited traits in offspring. Later on Weismann (1892) attributed a greater role to the influence of the environment, in the appearance of new hereditary changes. Hereditary germs or ids, in aggregates form chromosomes. At crossing there takes place an everlasting diversity of ids; hence Weismann's scheme of hereditary cell combinations.

At present Weismann's theory is only of historical interest. However, this explanation of reduction division and recombinations of hereditary germs is very close to modern conception of the processes of generative cell formation. Besides, Weismann's hypotheses on the essence of reduction division and of germ plasma made biologists deal with combinations and probability in respect to heredity phenomena.

In his " Experiments with plant hybrids", which was published in 1865 Mendel discussed the main regularities of inheritance and pointed in particular that;

1. Characters are determined by certain hereditary factors that are transferred through sex cells; and

2. Some characters of the organisms do not disappear but are preserved in the offspring in the same form as in the parents, though they may not always become manifest.

Mendel's theories are of great importance in understanding evolution. They reveal one of the sources of variability, namely, the mechanism of preservation of adaptive characters of species in a number of generations. If these adaptive

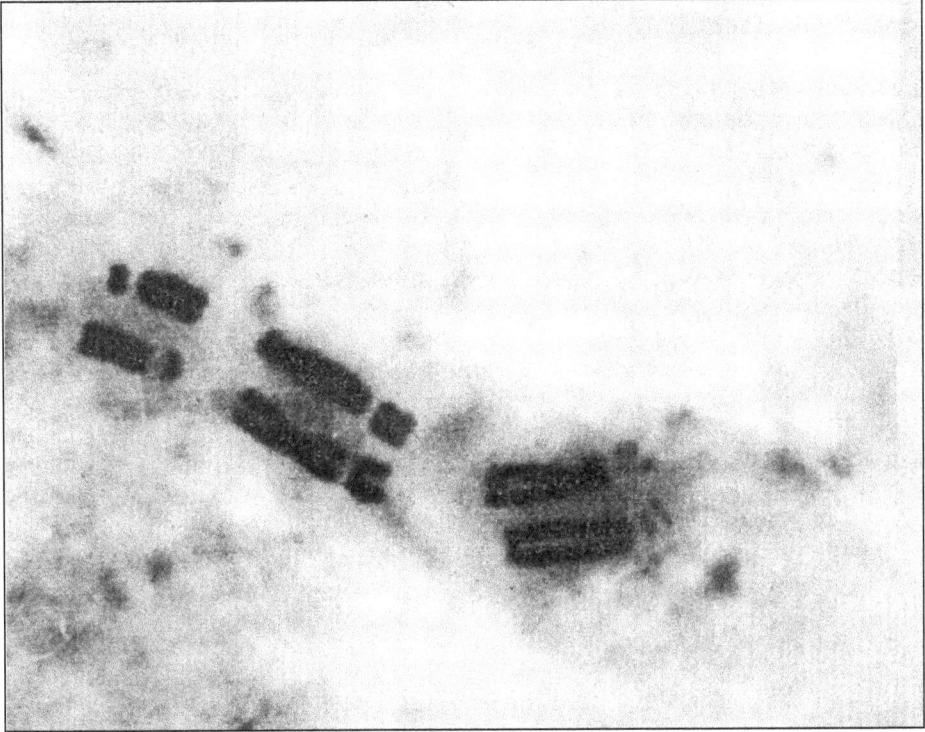

Figure 7.8: Normal Chromosome Set (Karyotype) of *Crepis capillaris* (2n=6).

characters were suppressed or disappeared at crossing, the development of the species would be impossible.

Mendel's regularities were reconfirmed by H. de Vries (1848-1935), C. correns (1864-1933) and E. Tschermak (1871-1962).

De Vries crossed mainly very close species distinctly differing in such obvious characters as flower colouration. He observed the phenomena of dominance and segregation in maize, poppy and many others. He published his observations in 1900, mentioning that the most significant points of these conceptions had been stated much earlier by Mendel in his work on the pea. They had been forgotten, however, and remained unrecognized. De Vries was the first to introduce such notions as monohybrids, dihybrids and polyhybrids.

Carl Correns (1864-1933) started his investigations of hybrids in 1894 and made a series of experiments crossing maize, peas, lilies and gilly flowers. In his article, published in 1900, on the results of hybridization experiments,Corrence first described the main principles of Mendel's discovery, whose paper he had become acquainted with in 1899. Then he presented his own data based only on crossing peas. It is of interest that Correnens visualized the possible existence of a cytological splitting mechanism. He mentions the sexual cell nucleus as the bearer of dominant and resessive dispositions and adds that the numerical ratio of gametes 1:1 suggests the idea of nuclear division, in other words, Weismann's reduction division.

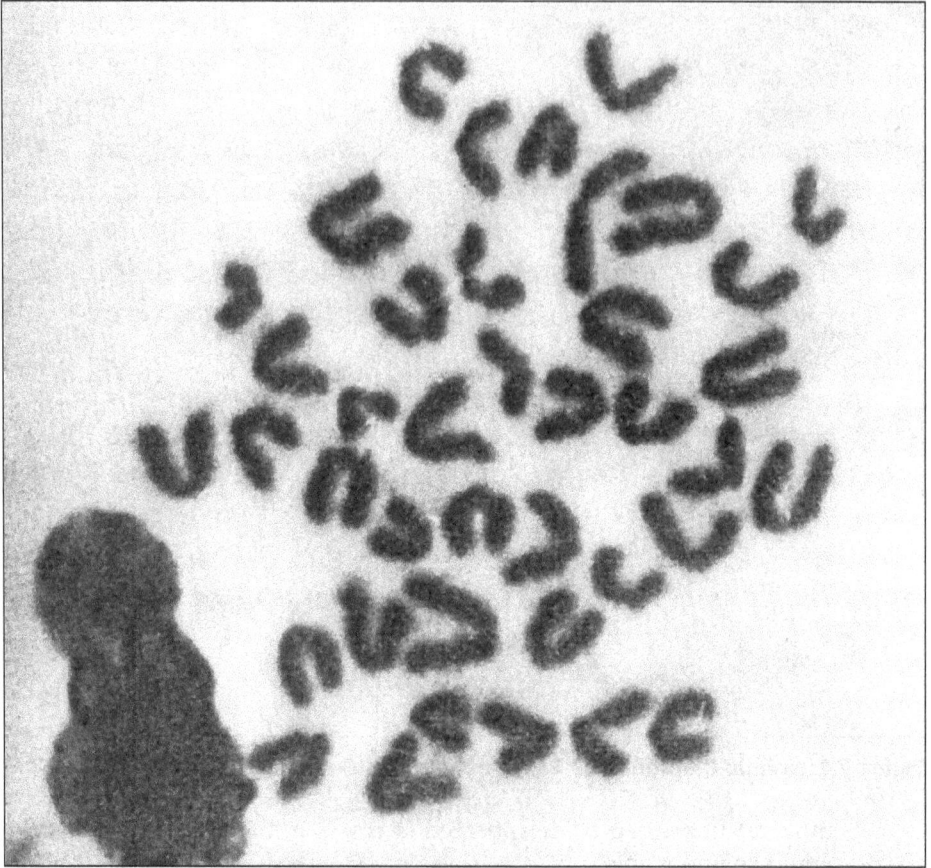

Figure 7.9: Normal Chromosome Set of Mice (2n=40).

In 1898 Eric Tschermak started experiments on hybridization of peas. In 1899 he discovered some regularities in peas, and shortly afterwards he came across Mendel's paper. In 1900 Tschermak presented his own paper for publication.

The discovery of hereditary regularities was immediately picked up by many scientists who devoted themselves to a new series of experiments that confirmed and developed those made by Mendel. It is evident that Mendel's discovery and the confirmation of his ideas was a natural result of research in the field of reproduction, fertilization and evolution, as well as variability and inheritance.

Mendelism and the Theory of Evolution

Mendelism played a decisive role in the elucidation of important aspects of Darwin's teaching and served as the basis of modern teaching on natural selection. But it was not realized that Mendel's ideas did not find due scientific recognition, and his first followers were attacked severely. The controversy went so far as to oppose Mendelism with Darwinism. Though inheritance as the basis of Mendelism does not entirely coincide with the theory of evolution which is the basis of the teaching of Darwin, they are nevertheless closely interrelated.

The controversy concerning Mendelian regularities stimulated scientific thought and resulted in the emergence of ideas and theories that, despite the controversy, played a certain role in the development of evolutionary theory and genetics. Most important of these was the theory of mutation of H. De Vries (1901-1903) which, according to the author's original ideas, completely solved the problem of species formation. The following concepts of H. De Vries' theory are still valid;

1. Mutations (considered mutation to be uneven change of a hereditary character) occur suddenly, without any transitions.
2. New forms are quite constant and stable.
3. Mutations, in contrast to non-inherited changes (fluctuations), do not form continuous ranges, do not group round the mode. Mutations are qualitative changes.
4. Mutations occur in different directions; they may be either useful or harmful.
5. Mutations can occur repeatedly.

De Vries held the theory of mutation against that of natural selection. He believed that mutation can give rise to new species; in this case no natural selection would be required. In fact, mutations are only a source of inherited changes, the latter being the material for selection. At present it is known that gene mutations are estimated by selection only in a genotype system.

It seems that there is considerable scope for significant work on mutation. The study has become the principal mechanism of gene apprehension and now we are on the verge of getting to know the mechanisms of mutations.

Development of Modern Genetics

The theories and experimental data formed the foundations of genetics in the first stages of its development. They nearly coincided with the second discovery of Mendel's laws, lagging behind them only a little, and confirming once again the adequacy of biology for the understanding of the new science of genetics. As soon as genetics became established as a science, biology made a stride, ceasing to be descriptive science and joining the ranks of exact natural sciences.

The second stage in the development of genetics started in the second decade of 20th century. The most important development of the period was the establishment of the American genetic school of chromosome theory of inheritance headed by T. H. Morgan, showing clearly that hereditary factors were related to chromosomes, and elucidating the role of chromosomes in sex determination as well as sex segregation (as 1:1). The chromosome theory of inheritance explained the exceptions to Mendel's laws and their universality for all organisms reproducing sexually.

It is at this stage of development the genetics found its application in selection. In 1922 N. I. Vavilov formulated the law of homologous rows in hereditary variability.

The essence of the law was that plant and animal species of related origin have similar rows of hereditary variability. Vavilov stated that;

1. Genetically close species and kinds are characterized by similar rows of hereditary variability with such regularity that knowing a number of forms within one species one can foresee that availability of parallel forms in other species and kinds. The closer the species and kinds are arranged in the general system, the greater is the resemblance in the rows of their variability.

2. Whole families of plants are generally characterized by a definite cycle of variability; hence regularities of close species and kinds in polymorphism make it possible to predict the availability in nature or in artificial selection, of respective forms by means of mutation, in-breeding or hybridization.

Applying this law, Vavilov defined the centers of origin of cultured plants to be where the largest variety of hereditary forms is collected. Bearing in mind hereditary variability of any species, one could fill in the missoing links of the species relative to the former on the basis of the law of homologous rows. The law of homologous rows is of great importance to breeding new kinds of plants and animal species.

Another event, the significance of which was realized to the full extent only much later (since forties in connection with radiation hazard) was the discovery of mutagenic effects of X-rays on fungi (in 1925 by G. A. Nadson and G. S. Filipov and on *Drosophila* in 1927 by H. j. Muller). To get an idea of the revolution that this discovery made in genetics, it is worthwhile recalling that genetic analysis is based on the study of mutations; that is, different forms of existence of one and the same gene. With the discovery of the mutagenic effect of radiation, there appeared a possibility of increasing yield by hundreds and thousands of times. While this discovery limited the use of theoretical experimental genetic analysis, to the applied genetics especially the selectionists, it gave a tool of considerable importance; that is, a number of mutations in the time unit, which made it possible to intensify selection work. These reports were the first to show that gene variability was affected by environmental factors. The study of the influence of ionizing radiation on inheritance resulted in the emergence of a most important branch, the radiation genetics. During the same period S. S. Chetverikov (USSR) starting elaborating theoretical bases of population genetics. In the thirties, V. V. Sakharov and M. S. Lobashev in the USSR and later S. Auerbach in England found hereditary changes influenced by chemical substances. In other words, chemical mutagenesis was discovered. It is of interest that the real progress in investigation on chemical mutagenesis could be witnessed only in the fifties, after the main concepts of gene structure had become available.

In the thirties a group of Soviet geneticists established some facts concerning the structure and function of genes. These findings anticipated the discovery of molecular genetics made a few years later and appeared to be the starting point of molecucular genetics of higher forms. The discovery of pseudoalleles by a group of geneticists led by A. B. Serebrovsky showed that the gene is a complex unit consisting of several parts. Data obtained showed the possibility of crossing-over within one and the same gene; that is, the possibility of gene division.

Another worker, B. N. Sidorov showed changes in dominating characters in respect to gene arrangement in the chromosome (it was convincing evidence of position effect) discovered by A. H. Sturttevant in 1926. Position effect is such a change of gene location in a chromosome which resulted in the change of this gene manifestation. But more important was the establishment of another fact, namely. Feed-back dislocation effect. If the arrangement of genes in a chromosome returned to the original state, there would be no observed changes of gene effect. These facts indicated the significance of space arrangement of genes providing for their activity.

The period since the forties to the present time can probably be called the period of molecular genetics. This period is characterized by studies employing a combination of genetic and biochemical analysis, as well as physico-chemical and mathematical methods. Owing to this as much as to the use of micro-organisms, findings were made on the chemical structure and activity of the hereditary factors, the genes. Deoxyribonucleic acid (DNA) appeared to be the messenger of hereditary information in all cellular forms, and the manifestation of gene activity is concerned with the synthesis of specific protein-enzymes.

Genetics as an Applied Science

In the role of genetics as an applied science, it can be differentiated in two aspects. On the one hand are theoretical problems and their solution. Of paramount importance are such problems as structure and reproduction of hereditary material, the problems of mutations (gene changes), interrelation of a gene and a character, interrelation of inheritance processes, variability and selection in evolution and the problem of gene interaction. It is difficult to predict at the present stage of development, the applied aspect of these, but the solution of the problems is necessary for the progress of genetics as science. On the other hand, there are practical problems that call for solution, such as ways and means of increasing productivity in live-stock raising and plant-breeding, the problem of hereditary diseases, protection from harmful physical and chemical hazards (namely, radiation) and so on.

The more extensive use of atomic energy raised the problems of radiation hazards. It is known thet radiation affects both body cells (somatic cells) and sexual cells (generative cells). Radiation results in changes in genes and chromosomes. There occur some kind of disturbance in somatic cells that brings about radiation disease and in some cases neoplasms, but they are limited to the same generation; whereas any changes in generative cells lead to mutations that may be inherited by the descendant generations. It has been established that under the influence there occurs a large number of lethal mutations, their frequency being dependent on the radiation dose. The effect of the doses must be summed up for proper estimation. It has been shown that the effect of regular small radiation doses results in accumulation of harmful (negative), mutations that can progressively increase hereditary lethality, malformations and other severe diseases, the hereditary abnormalities which will become manifest in future generations. Therefore, it is evident that radiation may present a great genetic hazard unless appropriate protection measures are taken.

In recent years, genetics has acquired even greater importance in solving medical problems. According to available estimates, about 0.03 percent of each generation of human populations be affected by various hereditary illnesses. These include such severe ailments as schizophrenia, cretinism, haemophilia, etc. The progress of medical cytogenetics within the last decade shows that several hereditary illnesses are due to structural disturbances of the chromosomes. It may now be possible to work out new methods for prevention and control of these diseases.

Genetics has greatly helped the pharmaceutical industry by artificially inducing hereditary changes in the micro-organisms which produce antibiotics. These mutations were induced by ultra-violet light, chemical reagents and X-rays. It was possible in this way to increase the yield of certain strains from 200 up to 5000 units, thus decreasing the production cost of antibiotics by 25 times.

The etiology of malignant tumors of somatic cells (cancerous) is still enigmatic, this being the reason for lack of effective ways to control the malady. Large number of specialists in the field consider the changes of hereditary apparatus of somatic cells to be responsible for the origin of cancer. This kind of change may result from mutation. The theory has been confirmed by a number of experimental findings, and one may hope that genetics will contribute substantially in the cause of cancer control.

Genetics in Relation to other Biological Sciences

The main task of biology is the study of living matter, and in this respect the role of genetics is especially important since it covers two basic phenomena- the inheritance and variability. These are related to reproduction, which in turn has a physiological and biological basis. Individual development is determined primarily by manifestation of gene action in the process of ontogenesis, which means that genetics has a direct relation to embryology, besides physiology and biochemistry.

In the study of taxonomy, genetic methods are often used. In fact, specificity of chromosome pattern has long become one of the systematic tests. At present, methods of molecular genetics are being worked out, for instance, complementary hybridization of nucleic acids in application to the problems of systematics and evolution. It is hoped in the near future these methods will provide means for establishing new evolutionary relationships between organisms.

The application of genetic methods has made it possible to solve a number of problems specific to such sciences as biochemistry, physiology and embryology in a new and a more efficient way. By making use of hereditary changes and mutations, one can switch off and on almost all physiological processes, interrupt biosynthesis of metabolites in the cell, interrupt morphogenesis and the like. The use of genetic methods in these sciences enables to obtain nearly any model necessary for investigation of one or another problem.

Genetics plays quite a special role in the teaching of evolution. Inheritance variability and selection are the main factors of evolution. The role genetics played in consolidating Darwin's ideas has already been known. Selection involves the effect of gene action, and it is owing to the evaluation of gene action expressed in characters and properties of the organism that selection of genes occurs, creating a most valuable system, a genotype. That is how natural selection operates to create a definite genotype of the organism. Thus genetics has unraveled the main factors and mechanisms of interrelationship in evolution-inheritance, variability and selection.

Thus, genetics is connected with all theoretical and applied biological sciences, including medicine and agriculture. Inheritance and variability is inherent in all nature, and this determines the position of genetics in the whole system of biological sciences.

Chapter 8

Hybridization

Hybrid Fish

Hybrid fish are not rare in nature. Slastenenko (1957) listed 212 known hybrids belonging to to 23 families, of which 91 are between cyprinids. Hubbs (1955) has shown that in suitable conditions, the rate of natural hybridization may be high. For example, where two allied species of fish more or less share the same spawning ground, but where one of the species are more abundant than the other, hybridization may occur because individual of scarcer species have difficulty in finding mates of their own species. In such conditions, Hubbs found upto 90 percent hybrids among *Lepomis* in Michigan stream. In fish culture, where a male and a female of different species can be isolated in a pond, breeding will often take place. So it is with several species of *Tilapia*; even intergeneric crosses may made in such conditions, as with members of the sturgeon family *Acipenser* and *Huso*.

The development and improvement of induced spawning by hormone injection has widened the possibilities of fish hybridization.

Objective of Hybridization of Fishes

The motives behind hybridization in fish is a search for improved qualities. Since many inter-specific crosses among higher animals are known to be sterile 'mules", inter-specific and inter-generic hybrids among fishes might prove sterile. In view of the great fecundity of fish, and the problems posed by the "wild breeding" of fish under cultivation, sterile fish would be valuable. With such fish planned crops would become generally feasible. Further, sterile fish might have a faster rate of growth or produce a fatter fish. They might have other useful qualities, such as, a non-migratory habit. But in fact, many fish hybrids, both inter-specific and even inter-generic, are not sterile. Sometimes, but not always, they have a faster rate of growth than the parent stocks, and they may have an improved shape.

A very much skewed sex ratio sometimes results from hybridization and in cases where hybridization gives 100 per cent of one sex, the hybrids are hypothetically as useful as sterile fish for stocking with the added advantage of population control.

Caution is needed in hybridization work. For example, it is possible to breed out good qualities as well as to produce them.This may especially be so where a fish has some specialized quality which makes it valuable. Examples are fish such as the grass carp, *Ctenopharyngodon idella* and silver carp, *Hypophthalmichthys molitrix*. The former has a pharyngeal dentition specialized for the rasping and grinding of fresh plant material; while the latter has a filtering apparatus on the gill arches which enables it to strain off feed or very small particles, such as, phytoplankton and including blue green algae. If the hybrids of these fishes with other cyprinids are without these useful characters, the hybrids will be less, not more, useful than the parent.

Further, too many hybridization experiments could endanger a fish breeding program, in as much as it might not always be possible to be sure that individual fish used as breeders are in fact from a pure strain of the species which it is desired to breed.

Difficulties in Hybridization Work

The large scale breeding of hybrid fish is in practice limited by the number of ponds available. Most fish breeding stations have pond space enough only for the breeding of known useful species, and it is the general experience that they could do it with much more.

Schaperclaus (1959) has pointed out, to follow a cross to the F2 and F3 generations would involve the rearing of very large numbers of fish to maturity. If such fish have to be divided among many ponds, qualities due to heredity may be masked by differences due to varying fertility of the ponds. The effect of environment may be greater than that of heredity, where fish have to be divided among many ponds, thus vitiating comparisons which might point to improved qualities of hybrid fish.

Fish hybridization work is very recent; and it seems that where it is being done, some, at least, of the hybrids should be reared to a size at which their properties, such as, colour, meristic characters, sex ratio, gonad development and rate of growth in comparison with and under the same environmental conditions as the parent strains can be defined. As yet there is little such information. Work can be concentrated on those hybrids, whether inter-specific or inter-generic, which show characters desirable in fish culture and superior to those of the parent strains and the rest can be rejected.

It would seem that the whole justification for setting aside pond space for the growing of hybrids is the practical one or producing fish with qualities superior to those of the pure strain now being bred, or which may be bred in the future.

Chapter 9

Hybridization in Fish Culture

Preferences for hybrids are based on inherent characteristics of hybrid heterosis (hybrid vigour) which manifest as intensive growth rate, higher viability, adaptive flexibility and sometimes early maturation. Charles Darwin attached great significance to hybridization. Even if not taken it for granted, at least it can be considered highly probable – the existence of a great law in nature, the law of crossing animals and plants which are not closely related to each other, and this is extremely useful and even necessary.

Hybridization has become a common practice in fish culture, though it has not attained its rightful place in the branch of economy. The hybridization of fish for practical purpose has demonstrated its economic value.

Natural Hybridization

Study of causes and regularities of hybridization processes occurring in nature reveals possibilities and methods of utilization of hybridization in fish culture. In no other group of the animal world, but fish do remote crossings (inter-specific and inter-generic) so frequently in nature, producing numerous hybrids. This is particularly true of fresh water fishes, as reproductive isolation is violated more frequently among them than among sea fishes, because of less constant hydrological conditions in rivers as compared to the seas.

Two hundred twelve different fish hybrids have been recorded up to 1956, of which only thirty are of sea fishes. The number of hybrids in nature may be much higher, but not all of them are known or identified correctly.

In nature, hybrids occur mostly as single specimen, but some times large scale hybridization may occur as a result of violation of reproductive conditions of the species. Abundance of individuals of one species occurring along with less number of other species, may result in an increase of inter-specific crossing. That is why the probability of crossing is high when large number of a species are introduced into a new area, while it remains low in the aboriginal species.

Fish hybrids are of interest for two reasons;

1. As object for commercial fish culture when heterosis of the first generation is to be used; sterile hybrids are suitable for this purpose, and

2. As initial material for producing new hybridogenic forms when the ability for reproduction is a necessary condition. The problem of hybrid reproduction is the most important problem in hybridization.

Great variations in the reproductive ability of hybrids have been observed; from complete fertility of inter-generic hybrids of some Cyprinodontidae to complete or incomplete fertility of inter-specific and inter-generic hybrids of sturgeon (Acipenseridae), the carp (Cyprinidae), sunfishes (Centrarchidae) and others.

The degree of fertility of hybrids does not always correspond to the degree of taxonomic relationship of the species crossed. In fact, inter-generic hybrids may be more fertile than intra-generic ones. Sometimes, even reciprocal hybrids produced by crossing the same species but by substitution of a female of one species by a male of another, and vice versa, may be different in fertility.

Individual hybrids may vary in fertility. In fact, some individuals of so-called sterile hybrids can produce offspring. Therefore, one of the method of eliminating complete sterility in such hybrids is to carry out hybridization on a larger scale to provide large number of hybrid individuals, some of which may be fertile. Fertility may be expected to increase in the second hybrid generation. This is true of such hybrids as *Cyprinus* X *Carassius*, which as a rule are sterile; but among them some exceptional individuals may be fertile. Investigations carried out shown possibilities of successful selection work on Ukrainian hybrids of *Cyprinus* X *Carassius*.

Fertility of most hybrids is disordered to a certain extent, and they are sometimes sterile. In the process of evolution of various species, they became gradually isolated ecologically, geographically and in other respects. Due to isolation a sort of barrier appeared which prevented them from free crossing. However, such an isolation was not always reliable; the possibility of inter-specific crossing still remained. This resulted in the development of a second barrier- disorder in hybrid fertility, preventing them from mixing and destroying specific differences. But where the first barrier acts without failure, fully eliminating the possibility of two species crossing, the second barrier might not appear. If such species are crossed artificially, the first barrier is removed, and a fertile hybrid is produced. Charles Darwin meant these very cases, when he writes, "there exist species which are not easy to cross, but their hybrids once produced, are highly fertile". Experiments have shown that man having succeeded in crossing such species, may find access to reproductive potentials that do not occur in nature.

Artificial (Man Made) Hybridization

It has been possible to produce fertile hybrids of beluga (*Huso huso*) and starlet (*Acipenser ruthenus*), two species which do not cross in nature due to marked ecological differences.

The sturgeons are remarkable for their ability to cross and for their hybrids that are fertile to various degrees. For many years sturgeon were crossed in most specific combinations in order to produce inter-specific and inter-generic hybrids, not only in the first generation, but also in the second and even the third generation (also back-cross and triple cross). In this work any instances were not noticed, where it was impossible to produce viable offspring.

Without detailed cytogenetic analysis of the causes of fertility and sterility of fish hybrids, there are more chances of producing completely fertile hybrids if the species caused have equal chromosome number which must be homologous and able to conjugate at the course of gametogenesis. When there is unequal chromosome number and violation of conjugation process, sterility occurs in various degrees.

The various species of sturgeon family such as, *Huso huso*, starlet (*Acipenser ruthenus0*, starred sturgeon (*A. stellatus*), ship sturgeon (*A. nudiventris*) and some others are characterized by a small number of chromosomes (about 60), whereas the chromosome number of *A. guldenstadti* is more than twice as much. Therefore, crossing of the first four species in any combination is likely to produce fertile hybrids; crossing each of the four with *A. guldenstadti* produces hybrids, whichare either of limited fertility or are completely sterile, the results of abnormal gametogenesis. Hence, before starting the work on hybridization, it is necessary to study chromosome complexes of the species to be crossed; which will permit prediction with a greater degree of possibility as to whether or not hybrids will be fertile or sterile.

Heredity of Hybrids

Many scientists, T. V. Mitchurin in particular, have shown that development of some characteristics of hybrids, to a great extent, depend on the environment in which the hybrids were bred. This environmental selection of characters is possible due to the great heterozygosity of the hybrid.

Study of morphological characteristics of certain hybrids of the sturgeons, the carps and others (as compared to their parent species) shows that hybrids have an intermediate heredity not only in the first generation, but also in the second, as well as in the generations produced by back-crossing and triple-crossing. In comparison to the first generation, the following generations, often but not always are characterized by greater variability; however, no typical morphological segregation followed by return to the initial species has been observed. So it was believed, that inter-specific hybrids possess the so-called permanent intermediate heredity which does not obey Mendel's law of segregation. But recent experimental research has proved that the type of heredity observed in crossing different species does obey the law of segregation. But the law, revealed by the behaviour of genes, is impossible to observe visually morphological characteristics. The species crossed are different in many characteristics, and consequently in a great number of genes. In inter-specific crossing high polymeric inheritance occur when numerous genes affect the development of some definite characteristics in a similar way. This is why in the second generation much greater number of combinations is inevitable; and among the mass of intermediate specimens only a few of them are homozygous, which in all characteristics show a return to the parental form. However, often such individuals are not found in the experiment.

In intra-specific crossing (inter-racial for example) it does not matter which race is represented by the male or female, since the characteristics of the offspring are not affected. That is to say, reciprocal hybrids do not differ from each other. This is expressed by the formula A X B = B X A. In inter-specific, inter-generic and more remote crossings, reciprocal hybrids may be different

Differences of reciprocal forms of some fish hybrids are revealed in the followings; (i) viability, (ii) growth rate, (iii) rates of development of other characteristics of heterosis and (iv) morphological characteristics.

Marked difference in viability has been observed in reciprocal inter-generic hybrids of Crucian carp (*Carassius carassius*), the offspring of *Carassius* females crossed with males of other genera of Cyprinidae (for example *Carassius* female X *Tinca* male) are viable; whereas the reciprocal offspring (of *Carassius* males) are not viable. Considerable differences in growth rates of fry were found in reciprocal hybrids of the sturgeons. The hybrid possessed some features, characteristic of the maternal species, revealing matroclinic inheritance. Morphologic differences between reciprocal hybrids are even more more pronounced in countable characteristics (number of vertebrae, fin rays), these differences are again mattrilinear.

Differences of reciprocal forms of hybrids are determined by cytoplasmic differences of the species crossed, resulting in different interaction of the nucleus and the plasma in reciprocal crossing. Genetic predetermination of cytoplasm may lead to incompatibility of the cytoplasm in certain species with chromosomes of some other species. As a consequence, some reciprocal hybrids may be viable, whereas some others may not be. The same is true of the factor determining the manifestation of various degrees of heterosis in reciprocal hybrids. Cytoplasmic heredity is a basis of matroclinic inheritance as well.

The nature of many phenomena in remote hybridization can not be explained without consideration of gynogenesis as one fundamental problem of the theory. Gynogenesis is a form of sexual reproduction, when in the process of fertilization, the sperm penetrates the ovum and activates it for development without contribution of paternal chromosomes, so that heredity in the offspring is determined by the female pronucleus alone. As a result, pure matroclinal pseudo hybrids are produced, represented as a rule by females only. Some scientists consider gynogenesis to be a special case of parthenogenesis. This is not correct because in parthenogenesis, activation of the ovum is not initiated by sperm but by some other agents.

Gynogenesis occurs in nature and can be demonstrated experimentally. Instances of natural gynogenesis are *Carassius auratus gibelio* and the Mexican viviparous fish *Mollienesia formosa*. They reproduce gynogenetically, spawning with males of some other species.

Some experiments on remote hybridization have resulted in so-called hybrid gynogenesis. Numerous experiments in crossing *Carassius carassius* females with *Leuciscus cephalus* males (different sub-families) have shown that only a small part of the offspring was viable, and these were matrilinear, that is, gynogenetic due to the exclusion of the male chromosome complex during development. Most of the embryos died at the very beginning of the development owing to incompatibility of maternal and paternal hereditary substance.

Of great theoretical and practical interest, is radiation gynogenesis, produced when eggs are fertilized by sperm exposed to X-rays in degrees sufficient to stop its hereditary function, but not its ability to penetrate the egg and activate its development. In fact, in the fertilization of starlet (*Acipenser ruthenus*) eggs by the sperm of the sturgeon exposed to X-rays, or the, beluga (*Huso huso*) eggs by the irradiated sperm of starlet, completely matrilinear offspring were produced (no parental characteristics were observed). Since gynogenesis produces only females, it would be possible to utilize it for control of sexual ratio in fishes, which is of great practical importance. Increase in the number of females of the sturgeon, for example, would be desirable both for commercial fishery and for artificial culture.

Heterosis in Nature

Heterosis in fish hybrids is a natural phenomenon observed in various families, such as, Acipenseridae, Salmonidae, Cyprinidae, Percidae, Centrarchidae and Poecilidae. As a rule, heterosis is more prominent in inter-generic hybrids which are not closely related, than in inter-specific or intra-specific hybrids.

Heterosis is revealed in various degrees. For example, growth rate of the hybrid may exceed that of parent species (typical heterosis) or sometimes growth rate of the hybrid may exceed that one of the parent species or be equal or less than the other parent (incomplete heterosis).

In the second hybrid generation, heterosis usually declines, but in back crosses of the first generation hybrid with the parent species (in which the characteristics that are of interest is more pronounced) the hybrid may exceed both the parent species, that is, the hybrid will show a prominent heterosis. For example, the hybrid of *Huso huso* X (*Huso huso* X *Acipenser ruthenus*) exceeds *Huso huso* in growth rate.

Both in evolution and selection, hybrid forms are of more value if their fertility is not violated.

Heterosis should be used on a greater scale, in selectional culture of new valuable fish and in acclimatization as well as in crossing of hybrids for commercial culture in ponds and other inland reservoirs.

Heterosis

Heterosis is superiority of a hybrid over the parent forms in the rate of development of one or more characters. The extent of development of a character is the result of development, characterizing this process quantitatively. Deviations in the extent of development of a character in hybrids may result in increase or decrease in manifestation, which are termed positive or negative heterosis. They supplement each other, providing an explanation to some effects of hybrid vigor. Both theories proceed from a common initial basis, namely hybrid vigor, which is the result of gene interactions. Interactions may be allelic and non-allelic. This result may be a consequence of interaction of separate genes as well as of gene groups. The heterosis effect of gene groups has been studied in populations of *Drosophila*. As a result of some structural modifications in chromosomes, such complexes of genes may appear which are not separated by crossing-over. These groups supplement each other when joining in a heterozygote, providing for better ability of heterozygotes as compared to the corresponding homozygotes. It may be concluded that natural selection in a population develops in such a way that gemets joining in pollination supplement each other in the best way and produce a generation of stronger constitution.

The scheme of the development of heterosis is to obtain a combination of lines which have the desired properties. The combining value of this line is estimated by means of crossing with another line of an adequately heterozygous population. The hybrids F1 produced by one inter-linear crossings very often cross between themselves. Selection of forms for crossing with the purpose of obtaining hybrid power is still based mainly on empirical matching of pairs. Development of general theory of heterosis could help in planning the experiments on crossing in such a way that the effect desired could be foreseen. Development of general theory of heterosis is retarded by the fact that all the details of gene interaction is not known.

Practical Application of Heterosis

The phenomenon of heterosis has been observed in all the species studied. In many cases, heterosis is so obvious that there is no need to resort to statistical analysis to demonstrate its value. This is particularly true in corn hybrids.

Methods of Inducing Heterosis

When two hybrid lines of corn are crossed, very often hybrids F1 produce twice as much seed yield as parents. The use of hybrid seed is now the main method of growing corn for grain and for silage. In order to produce hybreed seed, inbred (genetic nature of inbreeding in this case is the process of segregation of population in line with different genotypes) lines of good varieties are obtained which meet the requirements of the given climatic region (an inbred line is created for 5-6 years by means of random pollination).in selection of lines, their qualities are estimated in connection with the properties which should be obtained in a future hybrid organism. Inbreeding can not be effective if not accompanied by selection. Having created a great number of lines, crossing is begun. Inter-linear hybrids of the first generation are estimated by a heterosis effect; proceeding from this, the lines of better combining ability (combining ability of a line or species is heterosis development in hybrids obtained from their crossing) are selected and then reproduced at a greater rate for production of hybrid seed. At selection stations, work on production of inbred lines and estimation of their combining ability is carried out continuously. The more valuable lines are created as soon as it is possible to select better hybrid matching with necessary combination of characters. At present double hybrids of corn are used. The process of matching common hybrids for the purpose of obtaining the most productive double hybrids is a very important stage in the process of selection. The best results are achieved in crossing the lines originating from different varieties or strains. It is first necessary to investigate how long inbreeding should be carried on in order to achieve homozygosity by the group of genes, which are of interest to us, and second, to develop methods for rapid estimation of their combining ability. Crossing of inbred lines from one or various breeds has been widely used now in the field of poultry and pig breeding.

It is clear that in most cases, the inbred lines will always have lower indexes than the strains. Heterosis is evident only when the inter-linear hybrid exceeds not only its parent lines but also the varieties or breeds from which these lines generate.

The application of heterosis in fisheries is rather promising. High fecundity in fishes allows a considerable increase of yield to obtain adequate lines for crossing. A theoretical possibility of retention of heterosis in fishes with the aid of gynogenesis opens wide perspectives of commercial heterosis in fisheries. However, stress must be given on the wise use of hybrids in fisheries if only at high levels of pedigree stock farming and with valuable breeds are resorted too.

For fisheries, the combination of hybrid power and sterility can be of potential value. Tropical *Tilapia* can be taken as an example. This is a fresh water fish, and it produces so rapidly that over-population is often observed in ponds. In a over-populated pond a fish gains about 230 gram per year. However, some inter-linear hybrids develop powerful rapidly growing sterile males. In the ponds populated with such hybrids there is no over-population, the speed of growth, determined by heterosis reaches its complete development and the annual gain is 1360 gram.

Chapter 10

Hybridization of Fish in India–Asia

The common carp, *Cyprinus carpio* has been fully domesticated by selective breeding and a large number of strains and varieties have been produced in course of time. But the Indian and Chinese carps though have been cultured for thousands of years, these fishes have not been domesticated, and differ very little from their wild ancestors. This was possible because reproduction of common carp could be controlled, since the common carp reproduce in ponds and in confined water bodies. Since the Indian and Chinese carps do not breed in confined waters and only breeds in flowing river waters, this inability to breed under captivity has restricted attempts at selective breeding and hybridization of these fishes.

Success in induced spawning of Indian carps by injection of fish pituitary hormones was attained in 1957 (Chaudhuri and Alikunhi, 1957) and experiments in hybridization of Indian carps was initiated in 1958 by artificial fertilization of the stripped eggs of an injected female of one species by the sperms of an injected male of another species (Chaudhuri, 1959, 1961). Subsequent attempts were made to hybridize important species of Chinese carps and also the Indian X Chinese carps (Alikunhi *et al.*, 1963).

Fish Hybridization in India

Indian Major Carps

Successful hybridization of Indian major carps (Family Cyprinidae, sub-family Cyprinini) have been carried out by the following crossing method.

Fully mature female carps are given a preliminary dose of 2-3 mg of carp pituitary gland per kg body-weight. After an interval of six hours a second injection of higher dose, 5.8 mg/kg body-weight is given to each female. The male breeders are also given an injection of a low dose of 2-3 mg/kg body-weight. Depending on the temperature of water, within 4-6 hours after the second injection the female fish was found to be in oozing condition. The females and males are then stripped and the eggs and milt from two different species are mixed in a shallow container. Dry method of stripping is usually followed. Sometimes a little water is mixed and the tray is constantly tilted side ways. After a few minutes pond water is mixed and slowly the excess of milt is washed out. After several changes of water the eggs are transferred into biggertrays filled with fresh water. The eggs are fertilized and the blastomere appears on the animal pole within 45 minutes after stripping at a temperature range of 26 to 31 degree Celsius, and hatching takes place within 15-18 hours

The cultivated carps of India belong to three genera, *Catla*, *Labeo* and *Cirrhina*. Crosses and reciprocal crosses between eight species belonging to these three genera were carried out with success.

Inter-Specific Hybrids

Inter-specific hybrids were obtained by crossing four species, *Labeo rohita* (rohu), *Labeo calbasu* (calbasu), *Labeo bata* (bata), and *Labeo gonius* (gonius) in the following combinations.

Sl.No.	Parent Fish Species		Hybrid
	Male	Female	
1.	Labeo rohita	Labeo calbasu	rohu calbasu
2.	Labeo calbasu	Labeo rohita	calbasu rohu
3.	Labeo bata	Labeo rohita	bata rohu
4.	Labeo bata	Labeo calbasu	bata calbasu
5.	Labeo calbasu	Labeo gonius	calbasu gonius

In crossing the hybrids rohu calbasu and calbasu rohu over 94 percent fertilization was observed. Hybrids are extremely variable, though the general appearance seems to be like the paternal species. The growth rates of hybrids of both the reciprocal crosses were far superior to the slower growing maternal species *Labeo calbasu*.

The embryonic development of the three hybrids bata rohu, bata calbasu, and calbasu gonius was normal but the percentage of hatching was poor. Since *Labeo bata* is a medium sized carp, the hybrids bata rohu and bata calbasu did not show much promise although initially bata calbasu showed excellent growth.

The hybrid rohu calbasu was observed to have attained sexual maturity in two years. One pair of the hybrids was induced to breed by injection of fish pituitary and the F2 generation was obtained. Artificial fertilization of eggs by stripping was not necessary. The hybrids bred normally after injections of pituitary. Percentage of fertilization was high and several thousand young ones were produced. The hybrids possessed varying characters intermediate between *Labeo rohita* and *Labeo calbasu* (Chaudhuri, 1960, Government of India, 1961).

Inter-Generic hybrids

The following inter-generic hybrids were produced by crossing six species of fish belonging to three genera.

Sl.No.	Parent Fish Species		Hybrid
	Male	Female	
1.	Catla catla	Labeo rohita	catla-rohu
2.	Catla catla	Labeo calbasu	catla-calbasu
3.	Catla catla	Cirrhina mrigala	catla-mrigal
4.	Labeo rohita	Cirrhina mrigala	rohu-mrigal
5.	Cirrhina mrigala	Labeo rohita	mrigal-rohu
6.	Cirrhina reba	Labeo rohita	reba-rohu
7.	Cirrhina reba	Labeo calbasu	reba-calbasu
8.	Cirrhina mrigala	Labeo calbasu	mrigal-calbasu

(1) Catla-rohu hybrids : About 60 percent of the eggs developed and hatching took place after 16 hours. Mortality of hatchlings was high. Growth of the hybrid fry was faster than pure *Labeo rohita* fry obtained from the same mother. The hybrid characters were intermediate to the parental species. Catla-rohu proved to be an improved variety having a smaller head than *Catla catla* and wider body than *Labeo rohita*. Its growth is faster than the latter species, although not as quick as the former. However, it has the advantage over catla in having a smaller head. Size to size, the quantity of flesh is more in the hybrid than in pure catla. The hybrid attained full sexual maturity in three years. Although ill-formed testis were observed in a few specimens and the presence of both testis and ovary was observed in a single specimen, a good number of males and females were found to be normal and fully mature. The fecundity seemed to be less than the parental species.

(2) Catla-calbasu; The development of eggs was somewhat abnormal at the beginning. Only 20 percent of the eggs hatched out and only a small percentage of them survived. The growth was satisfactory and the hybrid reached the adult stage.

(3) Catla-mrigal : Since freely oozing catla male was not available at the time of the experiment, only a few mrigal eggs cold be fertilized, of which only a few hatched out. The early growth was satisfactory but all the hatchlings died due to pollution of water while being reared in the laboratory.

(4) Rohu-mrigal; and (5) mrigal-rohu : In crossing reciprocal hybrids rohu-mrigal and mrigal-rohu over 90 percent eggs were fertilized. The majority of the body characteristics of the offspring were intermediate to those of the parents. Both the hybrids matured fully in two years.

Figure 10.1: A Haul of Genetically Superior Rohu, "Jayanti".

(5) Reba-rohu and (7) reba-calbasu : Twenty percent of the fertilized eggs hatched out but all of them died on the third day.

(8) Mrigal-calbasu : Sixty percent of the eggs hatched out. The hybrid had a slightly finged lower lip and two pairs of prominent black barbells resembling that of the maternal parent. Colour of the body was intermediate between the parent species. The hybrids attained sexual maturity in 2-3 years. One eight year old female was later bred by hormone injection.

The inter-generic hybrid catla-rohu attained full maturity in three years. Four year old fish were selected as breeders for the F2 generation. The female received two injections of fish pituitary hormones at an interval of 6 hours and the male a single dose. When released together, they bred within five hours. Percentages of fertilization and hatching were high. Thousands of young hybrids were reared in ursery ponds. Their growth rate, when stocked along with pure *Catla catla* and *Labeo rohita* fry was observed to be higher than the latter, but less than the former species. Hybrids were very variable in morphological characters. They grew to adult size but the maturity has not been studied.

Outcrossing of Inter-Generic Hybrids

One fully mature mrigal-calbasu female hybrid was injected with fish pituitary and stripped. The stripped eggs wer fertilized by the milt from males of *Catla catla*, *Labeo calbasu* and *Cirrhina mrigala* and the following hybrids were produced.

Sl.No.	Parent Fish Species		Hybrid
	Male	Female	
1.	*Catla catla*	Mrigal-calbasu	Catla-mrigal-calbasu
2.	*Labeo calbasu*	Mrigal-calbasu	calbasu-mrigal-calbasu
3.	*Cirrhina mrigala*	Mrigal-calbasu	Mrigal-mrigal-calbasu

A good percentage of number of eggs hatched out and the hybrid fry that survived have been reared in ponds. Some males have been observed to have matured in one year.

Hybridization of Indian Carps with Common Carp

It was possible to successfully fertilize stripped eggs of *Labeo rohita* with the milt of *Cyprinus carpio* (Alikunhi and Chaudhuri, 1959). The eggs of Indian major carps swell to the size of 4 to 6 mm in diameter, are non-adhesive and demersal. The incubation period is 15-18 hours at temperatures of 26 to 31 degree Celsius. The eggs of common carp on the other hand are very small with very little perivitelline space, are adhesive and the incubation period is at least 48 hours at the above temperature range.

Though the percentage of fertilization was satisfactory subsequent mortality of eggs was high. Many larvae died at the time of hatching or soon after hatching. Only a very small percentage of hatchlings could be grown to adult size. The hybrid has elongated dorsal fin like the parental species but the other characters are intermediate between the parent species. Adults did not mature.

Reciprocal crossing was also tried. Though the eggs of *C. carpio* could repeatedly be fertilized with the milt of *Labeo rohita*, the eggs did not hatch out.

An outcross with the female hybrid rohu-calbasu and male *Cyprinus carpio* was successfully carried. The hybrid has been reared in ponds.

Hybridization of Chinese Carps

Inter-generic hybridization between three species *Aristichthys nobilis* (big head), *Hypophthalmichthys molitrix* (silver carp) and *Ctenopharyngodon idella* (grass carp) was attempted. In a few cases the eggs got fertilized but did not hatch out, while in others the hatchlings survived up to a maximum of one week. The crosses were :

Sl.No.	Parent Fish Species	
	Male	Female
1.	Silver carp	big head
2.	Grass carp	big head

Hybridization of Indian Carps with Chinese Carps

The following different species of Indian and Chinese carp were crossed, but the hatchlings did ot survive beyond one week.

Sl.No.	Parent Fish Species	
	Male	Female
1.	Silver carp	rohu
2.	Silver carp	catla
3.	Grass carp	rohu
4.	Grass carp	mrigal
5.	Grass carp	catla
6.	Silver carp	catla
7.	Big head	catla
8.	Big head	rohu

Fish Hybridization in Japan

Japan has produced several varieties of gold fish (*Carassius auratus*) and common carp (*Cyprinus carpio*) by long-time selective breeding; and inter-specific and inter-generic hybridization of fish has been started recently. Major part of the work on hybridization in fishes has been carried out on the cyprinid and cobitid fishes. Series of experiments on the hybridization of the cyprinid fishes belonging to the sub-family, Gobiobotii, have been conducted with a view to elucidate whether or not the success of hybrid development is correlated with the supposed degree of phylogenetic relationship based on morphological studies (Suzuki, 1961, 1962, 1963).

The following crosses yielded hybrids :

Sl.No.	Parent Fish Species	
	Male	Female
1.	*Gnathopogon elongates elongates*	*Pseodorasbora parva*
2.	*Gnathopogon elongates elongates*	*Gnathopogon japonicus*
3.	*Pseudorasbora parva pumila*	*Biwia zezera*
4.	*Pseudogobio esccinus*	*Biwia zezera*
5.	*Pseudogobio esccinus*	*Gnathopogon elongates elongates*
6.	*Pseudorasbora parva pumila*	*Gnathopogon elongates elongates*
7.	*Biwia zezera*	*Gnathopogon elongates elongates*
8.	*Carassius carassius auratus*	*Gnathopogon elongates elongates*
9.	*Pseudogobio esccinus*	*Pseudorasbora parva*
10.	*Biwia zezera*	*Pseudorasbora parva*
11.	*Biwia zezera*	*Gnathopogon japonicus*

Of these, reciprocal crosses have been obtained in all, except the last. The majority of the hybrids could be reared until they reached adult stage. The survival rate was usually similar to those of the controls. The hybrids were typically intermediate, not only in morphological characters, but also in behaviour. The few hybrids that reached adult size were found to be sterile males.

Fish Hybridization in Other Asiatic Countries

Taiwan

Tang (1965) has reported successful hybridization of Chinese big head and silver carp.Details of morphological characters, growth rate or behaviour are not available.

Mayanmar

During 1966, while conducting fish breeding experiments, Chaudhuri tried to cross the medium sized carp *Labeo pangusia* with *Labeo rohita* and *Cyprinus carpio*. Stripped eggs from an injected *Labeo pangusia* female were fertilized with sperms of *Labeo rohita*. Similarly another lot of eggs were fertilized with common carp sperms. The cleavages and early development of eggs were normal. The embryo was formed without any observable deformity. But the development stopped suddenly at that stage and no eggs hatched out.

Malaysia

In the Malacca Experimental Station some hybridization work was carried out on *Tilapia* spp. Hickling (1960) experimented with *Tilapia mossambica* and demonstrated that when an African male was crossed with Malayan female the offspring were nearly all males.

Indonesia

In Indonesia several coloured varieties and strains of common carp have been produced by selective breeding.

Viability of Fertilized Eggs during Hybridization

Successful hybridization and viability of fertilized eggs depend on several cytogenetic factors, most of which could probably be explained by cytogenetic studies. While conducting hybridization works on carps, it has been observed that it is easier to hybridize fishes which are closely related to each other, especially within the same sub-family. Indian carps belong to the sub-family Cyprinini and the Chinese carps come under the sub-family Hypophthalmichthyini. Hybridization within the sub-family Cyprinini was highly successful, but when tried between the two sub-families, the desired results were not obtained. Suzuki (1962) succeeds in a number of crosses and reciprocal crosses with several genera and species of Cyprinids, all belonging to the sub-family Gobiobotiini.

Viability of fertilized eggs may also depend on the length of incubation of eggs and size difference of the eggs of parent species. The incubation period of eggs of the Indian major and medium-sized carps are more or less the same. At the same temperature range, The Chinese carps eggs also hatch out approximately within the same time. Hatching success was noticed in a majority of crosses involving Indian and Chinese carps.

In the attempts to hybridize common carp with Indian carps it was observed that the eggs,though fertilized, did not hatch when a female *C. carpio* was used. The reciprocal cross with a female Indian carp was comparatively better since a small percentage of eggs hatched. The common carp eggs take at least 30 hours more for hatching than the Indian carp eggs. This suggests that female selected from the species having shorter incubation period gives better results.

Similar observations had been made by Buss (1956) in his experiments with Salmonidae. The incubation period of rainbow trout eggs is approximately 10 days shorter than the brook trout or the brown trout eggs. Whenever, male rainbow trouts were used to fertilize the eggs of another species the hatchability was nil. In the reverse cross, however, a small percentage of eggs hatched out.

In the hybridization experiment with the common carp and Indian carp the big difference in size of eggs also could be responsible for poorer results. Buss (1956) had also found that the size of the egg was a limiting factor. Lake trout embryos are generally larger than the brook trout embryos. So whenever a brook trout female is crossed with a lake trout male the progeny become crippled in the caudal region because of the confinement of a normally larger embryo in the smaller brook trout eggs.

Morphogenetic disturbance is also sometimes responsible for high incidence of mortality in the developing embryos or hatchlings. The reason for the entire mortality of hatchlings obtained in crossing grass carp and mrigal may probably be attributed to this factor.

Chromosomes compatibility is another limiting factor for successful species hybridization. The study of chromosomes as well as cytogenetic studies in the Indian carps have not been undertaken. It is of significance that in the breeding of the inter-specific and inter-generic hybrids of the Indian carps, the percentage of hatching of the F2 generation eggs was much higher than that of the F1 generation.

Fertility of Hybrids

In the hybridization of Indian carps all the inter-specific and most of the inter-generic hybrids obtained were fertile; and F2 generation of hybrids have been produced by crossing them. Rohu-calbasu, an inter-specific hybrid and catla-rohu, an inter-generic hybrid, both were bred successfully by hormone injections. Another inter-generic hybrid produced is mrigal-calbasu. Outcrossing by fertilizing its eggs with the milt of three species of carps belonging to three distinct genera were successful.

Experiments in Japan with Cyprinids gave varying results. Suzuki (1961) observed sterility in artificially produced inter-generic hybrids among bitterlings, but he also obtained an inter-generic hybrid of *Gnathopogon elongates elongates* and *Pseudorasbora parva*, the eggs of which were fertilized yielding F2 progeny. While he found the males of the inter-generic hybrid of *Pseudorasbora parva pumila* and *Gnathopogon elongates elongates*, producing abundant spermatozoa, the vast majority of his inter-specific and inter-generic hybrids were either sterile males or neuter. According to him, though these hybrid males had normally shaped testis, spermatogenesis was impaired. Observations made on the Indian carps showed that the majority of the inter-specific and inter-generic hybrids which grew to adult size attained full maturity although ill-formed gonads were present in a few adult hybrids.

Hickling (1966) in his review on fish hybridization has reviewed the work so far carried out on hybridization of warm-water pond fishes and the production of inter-specific and inter-generic hybrids.

Monosex Hybrids

In fish culture, very often the problem of over-population arises due to continuous breeding of some of the cultivated species. Since proper manipulation of stock is highly essential for obtaining higher production of fish in a pond, the fish culturists need to find out ways and means to prevent or control reproduction of these fishes. The problem is acute in countries where various species of *Tilapia* are cultured.*Tilapia mossambica* reproduces prolifically and soon over-populates the ponds, thus resulting in poor growth and poorer production of fish. Various methods namely, mono-sex culture by segregating the males, has been tried with initial success but failed to meet the desired results.

Inducing sterility or mono-sex production by cautious hybridization is an efficient method recently practiced for culturing the varios *Tilapia* species. Hickling (1960) got all male offspring in *Tilapia mossambica* from a cross of an African male (*T. hornorum zanzibarica*) with a Malayan female, which has considerable significance in tilapia culture. Since the hybrids are fertile the fish culturist should be cautious enough to remove the parent stock from the spawning ponds to prevent their mating with the offspring. Further experiments in this line in hybridizing various *Tilapia* species have resulted in the production of one hundred male progeny. Such hybrids were produced from the crosses of the following species.

Sl.No.	Parent Fish Species	
	Male	*Female*
1.	*Tilapia hornorum*	*Tilapia mossambica*
2.	*Tilapia hornorum*	*Tilapia nilotica*
3.	*Tilapia nilotica*	*Tilapia hornorum*

The phenomena of androgenesis and gynogenesis could suitably be applied in *Tilapia* to produce mono-sex offspring for culture.

Hybridization of Fish in Nature

Hybridization in nature among the Indian carps, though not very common, does occur. On rare occasions stray specimens of carp are met with, the external characters of which do not agree with any of the known species but are intermediate in characters between two species of carps. Such specimens are in all likelihood the hybrids produced in nature. The birbal fish, very often observed in the bundh type of ponds in the Midnapore and Bankura districts of West Bengal, India, where Indian major carps breed naturally during the monsoon months (Chaudhuri, 1966), is supposed to be a hybrid of *Catla catla* and *Labeo rohita*.

The conditions leading to hybridization of fish in nature are not known. When mature male and females of different species injected with pituitary failed to breed when put together. On few occasions the female ovulated but the eggs were not fertilized. Artificial fertilization of eggs with milt by stripping is essential for hybridizing defferent species of fishes.

In the bundh type of ponds where hundreds of Indian carps belonging to a number of species breed together in a small spawning area, sex-play between different species has not been observed so far. But occurrence of natural hybrids is fairly common in these ponds. The plausible reason for this seems to be the congestion in the spawning grounds, causing accidental fertilization of eggs by the sperms of another species.

Chapter 11
Fish Hybridization in North America

The Centrarchidae family (commonly known as sunfishes), a native of Mississippi river basin contains 27 extant species. Natural hybridization occurs quite commonly between certain species within certain tribes; however, natural inter-tribal hybridization is quiter are. Hybridization occurs frequently between the two *Pomoxis* sp. in the tribe Centrarchini and between many of the species within the tribe Lepomini. Largemouth bass (*Micropterus salmoides*) and spotted bass (*Micropterus punctulatus*) natural hybrids have recently been reported.

Although intertribal hybridization is rare in nature, laboratory experiments have revealed that viable hybrids can be produced from many intertribal crosses (West and Hester, 1966). Intertribal hybrids have been successfully produced between the centrarchini and the Lepomini and also between the Micropterini and the Lepomini. The white crappie, *Pomoxis annularis*, and the bluegill, *Lepomis macrochirus*, have been successfully hybridized. The warmouth, *Chaenobryttus bulosus*, and the largemouth bass, *Micropterus salmoides*, have been successfully hybridized both ways. The largemouth female has also been hybridized with the male bluegill, *Lepomis macrochirus*, and the F1 hybrid has a body shape that resembles the largemouth bass more than the bluegill. The green sunfish, *Lepomis cyanellus*, has also been successfully hybridized with the female largemouth bass. In most if not all of these intertribal crosses, there are partial lethals expressed and the numbers of deformed individuals are usually fairly high. Because of the rather typical expression of partial lethals in the intertribal crosses, much research must be conducted before the value of intertribal hybrids to fisheries management can be determined.

Hybridization

Species Used

Expression of partial lethals is much less frequent in intratribal crosses than in the intertribal crosses. The results of a study involving the hybridization of four species in the tribe Lepomini has been reported. The four species selected for study were the red-ear sunfish, *Lepomis microlophus*, the bluegill, *Lepomis macrochirus*, the green sunfish, *Lepomis cyanellus*, and the warmouth, *Chaenobryttus gulosus*. These species were studied because of local availability, taxonomic relationship, and similarities and differences in their morphology, habitat selection, and reproductive behaviour. All four species are sympatric in Illinois and there are broad overlapping of their periods of reproduction. The males of all four species construct saucer-shaped nests which they guard with great vigour. The males also protect and care for the eggs and larvae. When the larvae develop into free swimming fry they leave the nest and the males show them no additional parental care. The three *Lepomis* species nest in colonies, and the warmouth is a solitary nester. The colony nesters sometimes nest in mixed colonies; consequently the functional life spans of gametes could be very important in controlling hybridization between these species. If gametes are capable of fertilizing and being fertilized over long periods of time, sperm driftage could result in the production of hybrid individuals.

Experiments were conducted to determine the functional life spans of bluegill, green sunfish, and warmouth gametes. By stripping gametes and aging them for various periods of time prior to fertilization, it was determined that the average functional life span was approximately one minute for the spermatozoa and one hour for the ova. The brief functional life spans of the spermatozoa of these species are undoubtedly very important in reducing hybridization caused by sperm drifting from nest to nest.

Experiments Conducted

Two types of experiments were used to produce hybrid sunfishes. In the first, gametes were stripped from ripe adults and manually mixed. With this method it was possible to determine species isolation due to incompatibilities between sperms and eggs (primary genetic isolation). In the second method, designated isolation experiments, one or more pairs of fish composed of a male of one species and a female of another were isolated in small ponds to determine if they would hybridize when mates of their own species were absent.

Matings between individuals of different species are designated to P1 crosses and the resultant hybrids are designated as F1 hybrids. F2 hybrids are those produced by mating an F1 male with an F1 female. The P1 cross of a male bluegill with a female green sunfish is designated BxG and the resultant hybrids are designated as BG F1 hybrids; GB F1 designates the reciprocal hybrids.

Stripping Experiments

In the stripping experiments sperm and eggs stripped from the four parent species were paired in 16 different combinations to produce zygotes representing the four parent species and 12 hybrids. These experiments were designed to allow comparisons of rates of embryological development and the extent of viability of F1 hybrids and their maternal parent species.

Eleven stripping experiments were conducted. Eggs from the red-ear sunfish, three bluegills, three green sunfish, and two warmouths were fertilized with sperm from males of all four species. The temperatures at which these experiments were conducted were well within the range of temperatures that embryos of the four species are subjected to under natural conditions. In the following table the viability of each kind of hybrid is compared to that of its maternal parent species.

Table 11.1: The Degree of Viability of 16 different Kinds of Fishes Produced by Pairing Gametes from Red-Ear Sunfish, Bluegills, Green Sunfish, and Warmouths

Parent Species[1] Male x Female	Number of Eggs	Per cent[2] Hatched	Per cent[3] Normal Fry
R x R	512	39	27
B x R	512	41	33
G x R	552	46	37
W x R	548	32[4]	2[5]
B x B	681	87	75
R x B	742	86	83
G x B	639	87	69
W x B	699	61[4]	1[5]
G x G	639	78	75
R x G	597	80	79
B x G	589	73	70
W x G	678	76	55
W x W	295	58	49
R x W	317	62	58
B x W	311	58	44
G x W	276	62	47

1: R = red-ear sunfish, B = bluegill, G = green sunfish, W = warmouth; 2: Percentage based on number of eggs at the time sperm and eggs were mixed together and the number that hatched; 3: Percentage based on number of eggs at the time sperm and eggs were mixed together and the number of morphologically normal-appearing fry; 4: More than 90 per cent of these larvae were morphologically deformed; 5: These fry appeared morphologically normal, but all behaved abnormally.

No hybrid type was significantly different from its maternal parent species in the percentage of zygotes that hatched; however, more than 90 percent of the WR and WB F1 hybrids were morphologically abnormal.

Both WR and WB F1 hybrids exhibited high mortality between the hatching and swimming fry stages. At the closure of the experiments, only two percent of the WR hybrids and one percent of WB hybrids appeared to be morphologically normal. All of these morphologically normal-appearing WR and WB F1 hybrid fry were very sluggish, and it is very doubtful that any of these fry would have become free swimming. Fifty-five percent of the WG hybrids and 75 percent of the pure green sunfish zygotes developed into normal-appearing swim-up fry (difference significant to 0.05 level). The WG hybrid swim-up fry appeared to behaving normally. The remaining nine kinds of hybrids were not significantly different from their maternal parent species in the percentages that developed into normal swim-up fry.

The mean hatching time and standard deviation were calculated for each of the four parent species and each of the 12 kinds of hybrids. An analysis of variance revealed that WB F1 hybrid zygotes hatches significantly sooner than pure bluegill zygotes when both kinds of zygotes were incubated at the same temperatures. Although WB zygotes hatched in less time, the newly emerged WB F1 larvae were not as advanced in their development as the unhatched pure bluegill embryos. WR F1 hybrids were not significantly different from pure red-ears in hatching time; however, the newly emerged WR larvae were not as advanced in their development as the pure red-ear larvae. There were no statistically significant differences in the time of hatching between the either 10 kinds of hybrids and their respective maternal parent species, and differences in the degree of development between the hybrids and their respective maternal parent species were not pronounced.

The alpha temperature threshold of development (Shelford, 1927) and the mean number of developmental units (degree-hours of effective temperature) necessary for 50 percent hatching were calculated for 12 kinds of zygotes. It was impossible to calculate alpha temperature thresholds for zygotes involving warmouth females because of similarities of incubation temperatures. The alpha temperature threshold of development is the theoretical temperature below which normal embryonic development does not occur. T test comparisons revealed that the alpha thresholds of development of the 12 kinds of fishes were not significantly different from one another. The alpha thresholds ranged from 17.8 to 18.6 degree Celsius and the mean alpha threshold of all 12 kinds of fishes was 18.3 degree Celsius. Approximately 280 developmental units Celsius scale were necessary for 50 percent hatching.

Isolation Experiments

Thirty-two isolation experiments were conducted. Different species were isolated in small earthen ponds. Each of the 12 possible hybrid producing combinations was tested in one or more ponds. The R x G, G x B, and W x G pairing successfully hybridized each time they were tested.

The R x B cross was attempted in four ponds. No hybrids were produced in three ponds although the ponds remained full and were uncontaminated by other fishes. Eleven small fish were found when the fourth pond was drained, and these fish were believed to have been RB F1 hybrids although they were not positively identified as such (Childers and Bennett, 1961). The water in this pond contained a high and constant clay turbidity that reduced the transparency of the water and caused the parent fish to be extremely pale in body colour. The normally scarlet portions of the opercular tabs of the red-ear males appeared as a faint rose colour. The results of all other isolation experiments were either negative or inconclusive.

Fish hybridization might result from sperm driftage or interspecific matings. Sperm driftage is an important cause of hybridization among certain species of fishes, particularly minnows and darters which live in flowing water habitats and simultaneously spawn in close proximity to one another (Hubbs, 1955). Sperm driftage may also account for some hybridization between pond-dwelling or lake-dwelling centrarchids; however, since average functional life spans of sunfish spermatozoa are so brief and since there is such good synchronization in the release of sperm and eggs by a spawning pair, most hybrid sunfish are probably the result of interspecific pair formation.

The four experimental species are sexually dimorphic, closely allied, sympatric species. Signals that are in some way involved in reproductive isolation of such species are likely to be highly divergent and may involve specific differences in shape, colour, special movements, sounds, scents, etc. The precise signals which are operative in conspecific pair formation of the four experimental species are not known; however specific differences in colour of opercular tabs, eyes, cheeks, and pelvic fins of nest-guarding males may be important in controlling the behaviour of ripe females. When a female is ready to spawn approaches a nest-guarding male, she usually stops some distance from the nest and the male exhibits a courtship display (Miller, 1963). Species recognition apparently occurs during this short time, and the female flees or remains in the vicinity of the nest and accepts the advances of the male.

Since in one isolation experiment there was an indication that the scarlet portions of the opercular tabs of male-red-ear sunfish might possibly prevent hybridization between male red-ears and female bluegills, experiments were conducted to test this hypothesis. Four small earthen ponds were each stocked with three ripe adult male red-ear sunfish and three adult female bluegills. The opercular tabs were clipped from all males stocked in two ponds and the tabs were left intact on the males stocked in the two other ponds.

The ponds were drained during early October, and several thousand small hybrid fry were collected from the ponds containing red-ear males whose opercular tabs had been removed. No small fish were found in the control ponds. An examination of the clipped males revealed that the blue portion of the opercular tabs had regenerated to almost normal size but the scarlet portions had not regenerated. One tab on each of these males had a small, narrow, yellowish-orange margin.

Two such tests cannot, of course, be considered conclusive proof that specific differences in the colour of the opercular tabs of male red-ears are highly functional in preventing hybridization with female bluegills.

According to Hubbs (1955), fish hybridization is controlled to a large extent by environmental factors. Sunfish hybrids appear to be more common in ponds are choked with aquatic vegetation or have high turbidities than in clear-water ponds which have extensive spawning areas free from vegetation. In weed-choked ponds or ponds with high turbidities the range of visibility must be short, and under these conditions ripe females might occasionally spawn with males without observing preliminary courtship displays believed to be important in conspecific pair formation.

Hybrids Reared in Ponds

Large numbers of each of the 10 viable F1 hybrid types were stocked in one or more ponds. The F1 hybrids were reared to maturity in their respective ponds and the sex ratio, fecundity, and degree of heterosis of each F1 hybrid population were studied.

Sex Ratio of F1 Hybrids

Sexually mature F1 hybrids were collected from each population and sexed. Of the 10 kinds of viable F1 hybrids, seven were predominately males (RB, BR and BG were 97 per cent males; WG were 84 per cent males; and RG, GB, and BW were approximately 70 per cent males), two were approximately 50 per cent males (GR and RW), and one was predominately female (GW was 16 per cent males). Ricker (1948) determined the sex of 428 BR F1 hybrids in Indiana and found them to be 97.7 per cent males.

Sex determination in sunfishes is very poorly understood. Bluegills, green sunfish, and their F1 hybrids apparently have 24 pairs of chromosomes, and the sex chromosomes are indistinguishable from the autosomes (Bright, 1937). Bright also reported that the chromosomes are so similar in shape and size that he was unable to detect specific differences. Roberts (1964) found that red-ear, bluegill, and warmouth sunfishes each have 24 pairs of chromosomes; green sunfish from North Carolina had 24 pairs, but green sunfish from West Virginia had only 23 pairs.

The unbalanced phenotypic tertiary sex ratios of the F1 hybrid sunfish could result from unbalanced primary genetic sex ratios, specific differences in the strength of sex-determining factors, an overriding of the genetic sex by environmental factors, or differential mortality of the sexes.

Since the WG F1 hybrids were 84 per cent males and the reciprocal cross hybrids were 16 per cent males, it is possible that the strength of sex-determination factors of warmouths are 5.25 times more powerful than those of green sunfish. Specific differences in the strength of sex-determining factors cannot alone explain the sex ratios of the remaining eight kinds of viable hybrids, since none of these were predominately females.

RB and BG F1 hybrids were both 97 per cent males. If differential mortality were the cause of these unbalanced sex ratios, much of the mortality would have had to occur after the swim-up fry stages, since in the stripping experiments total mortality between fertilization and the swim-up fry stages was only 14 per cent for the RB and 27 per cent for the BG F1 hybrids.

It is not known which sex is the heterogametic condition for the sex chromosomes of the four experimental species; however, Haldane (1922) formulated a rule which furnishes a clue; "When in the F1 offspring of a cross between two animal species or races, one sex is absent, rare, or sterile, that sex is always the heterozygous sex." Using Haldane's rule, Krumhols (1950), in a study concerning

BR F1 hybrids, pointed out that the males of both bluegills and red-ear sunfish are probably homozygametic for sex and the females heterozygametic. The application of haldane's rule to all possible F1 hybrids produced from red-ear sunfish, bluegills, and green sunfish indicates that the female is the heterozygametic sex in these three species. Hybridization of male warmouths with females of the three *Lepomis* species resulted in partial or complete lethals, suggesting that in the warmouth the male is the heterogametic sex.

Reproductive Success of Hybrids

The reproductive success of each of the 10 kinds of viable F1 hybrids was investigated in one or more ponds. The occurrence and abundance of F2 hybrids were determined by seining, trapping, shocking, poisoning or draining the ponds after the F1 hybrids were one or more years of age. RB, BR, and BG failed to produce abundant F2 generations when in ponds which contained no other species of fishes. In contrast to these results, BR F1 hybrids produced abundant F2 generations in two ponds in Indiana (Ricker, 1948). The other seven kinds of F1 hybrids produced abundant F2 populations when stocked in ponds containing no other fishes. Three of the seven kinds of F1 hybrids which produced large F2 populations when stocked in ponds containing no other fishes were also stocked in ponds with largemouth bass. RG F1 hybrids and GB F1 hybrids, when stocked with largemouth bass, produced only a few F2 hybrids. No F2 hybrids were found in the pond stocked with BW F1 hybrids and largemouth bass. WG F2 hybrids and GW F2 hybrids were stocked in ponds containing no other fishes. Both of these F2 hybrids produced large F3 populations.

Backcrosses, outcrosses, a four-species cross, and a three-species cross involving F1 hybrids are listed in the table below. The BW x B backcross was made by stocking adult male BW F1 hybrids and adult female bluegills in a pond which contained no other fishes. The other 12 crosses listed in the table were made by stripping gamets from ripe adults and rearing the young to the free-swimming fry stage in the laboratory.

R x RW, W x RW, B x RW, G x RW, R x GB, and RB x W young were killed after they developed into free-swimming fry because of the lack of ponds in which they could be stocked. All six kinds of fry appeared to be normal and probably would have developed into adults. Free-swimming fry of the remaining six crosses in the laboratory were stocked in ponds and did develop into adult fishes. BW x B, G x GW, and B x RG populations produced large numbers of young.

Hubbs and Hubbs (1933) reported that in Michigan F1 hybrids of bluegills, green sunfish, longear sunfish, pumpkinseeds, and orange spotted sunfish were unable to reproduce because males were sterile and ova stripped from few adult females used in the experiments appeared distinctly abnormal. This study, often cited in the literature, has resulted in a rather widespread belief that all male hybrid sunfish are sterile. Results of the present experiments conclusively established that a number of different kinds of hybrid sunfishes produced in Illinois are not sterile, are fully capable of producing abundant F2 and F3 generations, and can be successfully back crossed to parent species and even outcrossed to nonparental species.

Table 11.2: Successful Backcrosses, Outcrosses, Four-Way Cross and another Cross Involving F1 Hybrid Sunfish[1]

Backcrosses *Male x Female*	Outcrosses *Male x Female*	4 Species Cross *Male x Female*	3 Species Cross *Male x Female*
R x RW	R x GB	RB x GW	BW x GW
G x GW	R x BW		
W x RW	R x GW		
BW x B	B x RG		
	R x RW		
	G x RW		
	RW x W		

1: R = red-ear sunfish, B = bluegill, G = green sunfish, W = warmouth.

Hybrid Vigour

Heterosis has been defined (Manwell, Baker and Childers, 1963) as " that condition where, with respect to one or more particular characteristics, the values for most, if not all, of the individual hybrids fall significantly outside the range formed from the means for both parent populations. In cases of positive heterosis – hybrid vigour – the hybrid shows a faster growth rate than either of the parents, or it possesses some other characteristic, often an economically significant one, at a 'better' level than the parents do."

Rate of Growth

In an attempt to determine whether certain F1 hybrid sunfishes actually grow faster than their parent species, two experiments were conducted in which equal numbers of uniformly sized F1 hybrids and parent species were stocked in ponds which contained no other fishes. Intraspecific competition is keener than interspecific competition because individuals of the same species are more nearly equal in their structural, functional, and behavioural adaptations. Consequently in experiments designed to compare rates of growth, it is imperative to use equal numbers of similarly sized fishes. In the first experiment the growth rate of BG F1 hybrids was compared to that of green sunfish. In the second experiment the growth rates of GR F1 hybrids, green sunfish, and red-ear sunfish were compared.

In both experiments the average increase in total length of the hybrids was not significantly different from the increases of the parent species. The population densities of the fishes in both ponds were much lower than would be found in most normal natural populations. In both experiments intraspecific and interspecific competition was undoubtedly quite light; consequently, the question of whether certain F1 hybrid sunfishes are superior to their parent species in rate of growth cannot be answered until high density populations containing equal numbers of equal sized hybrids and parent species are studied.

Some Current Studies in Cultured Fishes

The importance of North American work on hybridization and selective breeding of fishes probably lies not in previous efforts in this field, but in what is being done now and planned for the future.

The practice of fish culture itself is a notable selection of fish. Fish in their natural habitat are adapted to their environment as a result of physiological and morphological mechanisms acquired through millions of years of evolutionary influences. These adaptations developed slowly, except in times of catastrophic change or sudden mutation, but the impact of the fish cultural environment has been and is severe and rapid. Artificial selection may operate somewhat independent of the environment, but only briefly. Often it is thought that the hatchery environment is relatively safe and stable; it is not necessarily true, especially in the pond culture of warm-water species

The species now being cultivated around the world are newly introduced to fish cultural facilities, where environmental conditions are quite different from those in their natural habitat. Pond, lake, raceway or aquarium conditions impose different and usually severe stresses upon the fish. Only those with adaptive attributes and a viable genetic make-up survive and thrive. The intolerant individuals or populations either die or are discarded by the fish-culturist.

Breeding of Rainbow Trout

The breeding program was divided into three general areas; (i) identification of variants exhibiting simple Mendelian inheritance, (ii) development and maintenance of random bred and inbred stocks, strains or lines, and (iii) selective breeding for specific characteristics, such as, rapid growth and early maturity. The first is simply a matter of testing the potential Mendelian character by a series of crosses, backcrosses, intercrosses and incrosses. The second area involves selection of healthy individuals without regard to any special phenotypic character, from either unrelated or closely related stocks. The selective breeding phase requires most of the effort in the breeding program. The fish are selected primarily on the basis of performance in respect to the character under selection. However, other factors such as viability, especially in the early phases of evaluation, may decide if the lot is retained.

Two-way selection was practiced, mainly for control purposes. However, it is possible that fish possessing a character developed by downward selection could have intrinsic value, particularly for use in research laboratories.

The terms individual (often called mass), sibling, half-sibling and progeny testing are used to identify methods of selection. These terms simply indicate criteria used to evaluate the breeding worth of the mass lot. Individual selection refers to selection based solely on the individual's own performance or phenotypic value. In other methods, the broodstock is chosen on the basis of their relatives' performance, namely, the selection of potential breeders for DDT susceptibility on the basis of the response of their brothers and sisters to DDT exposure. Sibling, half-sibling, or progeny selection are forms of family selection, but the term "family" is also used where families (usually sibling lots) are retained or rejected as units in accordance with their mean phenotypic value. Here, the selected individual's own performances contribute to the selection. The method of selection is determined by the character of interest, availability of rearing space, and relationship of individuals in the group under selection. All these methods and a combination of methods are used when feasible.

Both factorial (often called diallel) and single matings are made for most purposes. In a given amount of space, the single mating approach provides greater potential diversification among the progeny since more parents are represented. However, the factorial design (each of a group of males mated to each of a group of females) is a useful tool for evaluating parental contribution as well as lots. A factorial mating produces a number of half-siblings for simultaneous evaluation. For example, a complete 4 x 4 factorial (4 pairs of parents) produces 16 sibling lots, and the fish in each lot are half-siblings to fish in 6 other lots.

A dry diet, open-formula granules and pellets, is fed in accordance with the premise that the feeding rate is inversely proportional to the cube root of the average weight of the fish (hatchery feeding charts reveal this relationship) The weight of daily rations for a given lot of fish equals the total weight of the fish load times a feeding factor or constant divided by the cube root of the average weight of the fish.

For example, employing a feeding constant of 0.1 (K), 10 kilograms of fish (Wt) averaging 8 grams in weight (Wa) are fed 0.50 kilogram per day; 10 kilograms of 64-gram fish are fed 0.25 kilogram. Using a slide rule for the calculations, this method provides a rapid means for determining the amount of feed without the aid of charts. A relatively high feeding factor was selected to prevent masking of genetic growth potential that might occur as a result of underfeeding. The use of this factor yields feeding levels that are essentially the same as those recommended by Duel *et al.*, in 1952.

Inheritance of Albinism

Albinism in rainbow trout appears to be a simple autosomal recessive. Several male and female albino and wild-type (normally coloured) fish were mated in all combinations. Albino x albino crosses produced all albino progeny; wild x albino and wild x wild resulted in all normally coloured fish. Progeny sibling lots derived from each type of cross have been reared to maturity for further test crosses. Recently, precocious F1 male progeny from wild x albino crosses were mated with F1 albino females. The backcrosses yielded albino and wild-type embryos in numbers closely approximating the 1:1 ratio expected in classic Mendelian inheritance.

Breeding for Genetic Diversity

This work involves the breeding and maintenance of random-bred stocks to preserve heterozygosity. Unrelated fish (individuals not known to be related) from 1964 or 1965 year class New Zealand, Sand Creek, De Smet, Donaldson, and Arlee albino stocks were mated during the 1967 breeding season to produce the random bred stocks for 1967. Similarly, broodfish from the above and 1966 Wytheville and

Manchester stocks are mated to yield current year class stocks. These year class stocks are being reared and maintained for future random breeding.

Breeding for Genetic Uniformity

The purpose of this work is to develop several different strains or lines characterized by genetic uniformity within lines and genetic diversity between lines. Sibling lots of New Zealand, Sand Creek, Donaldson, De Smet, and Arlee albino stocks were produced during 1967 breeding season. Mating activities for 1968 season have provided sibling lots of New Zealand and Sand Creek rainbows that are unrelated to 1967 lots. This year's mating also produced sibling lots of Manchester and Wytheville stocks. A few lots of each stock and year class are being reared to maturity for brother-sister inbreeding.

Selective Breeding for Year of Maturity

During the 1967 spawning season, trout that had matured in two years from SandCreek, New Zealand, and Donaldson stocks were mated. Also trout that matured in three years from De Smet, Sand Creek and New Zealand stocks were crossed with males of uncertain age at maturity from the same stocks. Approximately 30 lots are being reared to maturity for selection purposes.

Selective breeding for 1968 season was initiated with matings of two-year old maturing fish from Wytheville and Manchester stocks. Fish selected from 1965 Sand Creek, New Zealand, and Donaldson stocks that did not mature last year are being held for possible matings as three-year old (or older) maturing phenotypes.

Selective Breeding for Growth

Sibling lots were produced by random matings of Sand Creek, New Zealand, Donaldson, and De Smet stocks. The growth of 200 lots was measured for a period of 13 weeks following swim-up. The average weight attained by the best growing lot was about eight times greater than that of the poorest. Seventy lots were retained for further evaluation. Subsequent selection reduced the number of lots to 24 by the end of the year. The eightfold maximum difference in average weight has remained relatively constant since selection. Most of the remaining lots will be reared for another year.

Production of 1968 sibling lots has been underway since September. Broodfish from several stocks reared under similar conditions were individually selected on the best of attained size and have been mated. Evaluation of the early growth of approximately 100 lots has been initiated.

The substantial variation observed between the average weights of sibling lots and in the individual sizes of broodfish is very encouraging. It suggests a considerable amount of genetic diversity which, of course, must be present for successful selective breeding.

Selective Breeding for Tolerance of Crowded Rearing Facilities

Randomly selected groups of fish from stocks of 1965 Donaldson and 1966 Manchester, New Zealand, Sand Creek and 1967 Manchester rainbows have been reared in heavily loaded circular tanks. The average growth of fish in these tanks is reduced significantly and some mortality occurs. The fish will be maintained in these conditions as long as space is available. Surviving individuals exhibiting extremes of growth will be selected for breeding. The 1966 Manchester stock was the only group containing mature fish before the end of the year. About 50 matings of selected individuals have been made.

Selective Breeding for Formalin Tolerance

Randomly selected brood fish from 1964 and 1965 stocks yielded a number of 1967 sibling lots for evaluation. Sample fish from 170 lots were exposed at 14 weeks after swim-up to formalin concentrations of 175 and 525 microliters per liter for 6 hours. Resistant lots suffered less than 50 percent mortality at the high concentration while the low concentration killed more than 50 percent of the fish in susceptible test lots. Siblings or survivors from the more resistant and susceptible lots have been retained for future breeding.

Breeders for 1968 have been selected and mated in accordance with their own or their progeny's response to formalin. Also mating of unselected broodstock is producing additional sibling lots for evaluation. By the end of the year a total of 200 sibling lots was on hand, scheduled for testing in the spring.

Selective Breeding for DDT Tolerance

The 170 sibling lots of the 1967 year class evaluated for formalin tolerance were also tested for their response to DDT. Randomly selected test fish were exposed to DDT concentrations of 13.3 and 40 microliters per liter for 12 hours. The difference in tolerance of susceptible and resistant lots were similar to that noted for formalin. Generally, DDT-resistant fish were from lots that exhibited relatively poor formalin resistance. Selected lots are being reared for future breeding.

Here, as in the case of formalin, groups of the 1968 breeders are being selected on the basis of individual or progeny performance. Sibling lots produced by these parents and additional lots from unselected parents will be evaluated in the spring.

Warm-water Fish

Catfish Hybridization

The importance of hybrid fishes have been recognized for food and sport. (Smith, 1961). In many instances, hybrid animals possess desirable characteristics, such as, rapid growth, tolerance to unfavorable environmental conditions, resistance to disease, and other attributes not always apparent in either parent species. At the same time, a hybrid animal may not possess undesirable characteristics found in one or both of its parents. Catfishes are increasing in importance in warm-water areas of the world. Since hybrid catfish appeared to be of value in fish husbandry and management in the United States, research has begun to develop methods to produce and culture the hybrid catfishes and to make preliminary observations on their growth and desirability as compared to the parent species.

During April to mid-June, the normal spawning season for catfishes in south central United States, gravid channel catfish, *Ictalurus punctatus*, white catfish, *Ictalurus catus*, blue catfish, *Ictalurus furcatus*, flathead catfish, *Pylodictus olivaris*, black bullhead catfish, *Ictalurus melas*, yellow bullhead catfish, *Ictalurus natalis*, and brown bullhead catfish, *Ictalurus nebulosus*, were paired with sexually mature males of the same species and approximately equal size. Females were injected with fish pituitary or human chorionic gonadotropin to induce spawnibg. After spawning had commenced the female fish was netted, subdued with hand pressure around the caudal peduncle, and then held in a normal upright position, tail slightly below the head. By a gentle stroking of the abdomen, the eggs were stripped into a water-filled, plastic-film-lined basket. At the same time, small portions of excised, macerated testis from another species of catfish were dabbed into the water above the extruded eggs to bring about fertilization. After the desired number of fertilized eggs were obtained, the egg mass was left undisturbed for approximately ten minutes to water-harden. The water-hardened egg mass was then detached from the plastic film, placed in a woven-wire enclosure and positioned in a mechanical hatching device. Care of the eggs during embryonic development, hatching and yolk-sac period, and initial feeding stages of hybrid fry were similar to published techniques for channel catfish.

These techniques were used to produce 24 different hybrid types from 7 common species of freshwater catfish. Attempts to produce hybrids from eggs of flathead catfish and sperm from any of the six species of catfish resulted in infertile eggs or eggs which disintegrated a few days after apparently being fertilized. But the sperm from flathead catfish was used successfully to fertilize the eggs of the other catfishes.

Comparative Growth of Catfish Fingerlings

Three experiments were conducted in aquaria over a three-year period starting in June, 1965. Test fish for each experiment were hatched over a five-day period and when they stocked averaged 0.5 to 1.5 g each. In experiment 1 all categories of fish were offered equal amount of feed; those in experiment 2 were offered equal rates of feed; and those in experiment 3 were offered ad libitum amounts of feed. Fish were re-weighed at 3-week intervals and revised feed allowances calculated. At the termination of the 9-week experiments, weight gain and feed conversions were calculated.

Results of three experiments that compared six species of catfish with their hybrids on the basis of weight gain, growth potential and feed conversion indicate that some of the hybrids have possibilities as sport or food fishes. Also by utilizing selection and genetic studies, it can be speculated that growth rate and feed efficiency can be further increased. However, all factors considered, it appears even without selection and genetic studies that the obviously fast growth rate and efficient feed utilization of the white catfish x channel catfish hybrid make it a desirable fish for intensive culture. It has been observed in research ponds that such hybrid fish weighing weighing up to several pounds are more robust than the channel catfish and lack the large bull-head of the white catfish. It appears from some of the preliminary work that this hybrid has a greater tolerance to low oxygen than either the channel catfish or white catfish species. This in itself would make this fish desirable in intensive cultures since low level of dissolved oxygen associated with the large feed allowances represent the first limiting factor in production and the greatest threat of mortality.

It has also been observed that some catfish species and their hybrids perform differently in aquaria than in ponds. Possibly some of this difference is due to natural behaviour, that is, sensitivity to unnatural light and sounds or limitation of space inherent in aquarium facilities. A notable example is the blue catfish which competes actively with channel catfish in managed ponds and in natural lakes and rivers but performs poorly in aquarium experiments. Others of these hybrid groups which perform poorly in aquarium studies may perform well in managed ponds. Thus, before any hybrid can be discounted for use in sport-food cultures, preliminary tests should be carried out in ponds incorporating variables such as stocking and feeding rates and rates of water exchange.

Chapter 12

Commercial Hybridization and its Practical Significance

The application of hybridization in fish culture is closely connected with two major aspects. Firstly, it is a direct use of the results of crossing, that is, the effect of heterosis and favorable combination of some valuable features of parental forms in hybrids of first generation (so-called commercial or utilizable crossing). Secondly, and a no less important point, the prolific hybrid forms can represent a valuable material for further selection (so-called synthetic selection).

It is to be noted that in pisciculture, the use of heterosis is of growing significanc. Fish culture is primarily concerned with hybrids which show in the most conspicuous way the merits of the hybrid organism; that is, higher viability, accelerated rate of growth, and greater adaptability.

It has to be emphasized, nevertheless, that hybridization does not always lead to the improvement of qualities in offspring. The results of crossings can be quite variable and are conditioned by numerous factors. The results depend on the biology of parental lines (including their morphological and karyological characteristics), magnitudes and genetic structure of populations of initial groups, degree of remoteness, and the level of heterozygosity of the crossing forms. The success of crossings will be determined to a great extent by the knowledge of these particulars and a skillful selection of groups for obtaining a maximum possible effect of heterosis from crossing.

Therefore, it appears very important to have a clear idea about the nature of heterosis, the mechanism of selective action and a relationship between genotypic and paratypic variations.

Heterosis

It appears essential to define heterosis more precisely. It is generally understood by this term that the hybrid is superior in comparison to the parents in development or in the expression of other valuable features. Usually hybrids show accelerated growth and development, higher viability and greater resistance to unfavorable influence of environment and diseases. In natural populations, where the most important single criterion of an individual's value is its capability to leave a viable progeny (fitness), these manifestations of heterosis are of selective importance.

Among with occurrence of properties which raise the selective value of hybrids, it can be seen rather frequently the augmentation of size or of development in general, of various organs with lower adjustability and survival. The latter case is named by Dobzhansky (1952) as luxuriance (prosperity or gigantism) in contrast to heterosis proper (euheterosis) which is expressed in higher vitality and thus in higher selective value of hybrids.

In both cases the use of heterosis proves to be advantageous and can contribute to increase in productivity of animals and plants. It has to be taken into account, however, that specific features of display of heterosis can vary in species which have been cultivated by man for a long time. For instance, in the case of domestic animals and cultured plants, due to replacement of natural by artificial selection, the heterosis effect loses its selective character. Instead of a general increase in viability and fitness, its main feature can be expressed in acceleration of growth or even increase in size of individual organs and intensification of certain functions gradually in the second and following generations (Kirpichnikov, 1967). Therefore, a further and more complicated task of a selectionist is to fix the heterosis effect achieved in the first generation. The solution of this task can be feasible only in cases where the nature and origin of hybrid vigour are known.

Heterosis has been found in almost all living organisms from yeast to higher plants and mammals. Although widespread among organisms, heterosis shows a great similarity in different species. This leads to the assumption that there are some peculiarities common to all organic forms which cause heterosis. It seems advisable therefore to give some basic data related to the nature of heterosis or hybrid vigour, even though they are based on experiments with various species of plants and animals other than fishes.

Theories of Heterosis

To clarify the genetic mechanism of heterosis, two hypotheses are usually considered; the hypothesis of combination in hybrids of favorable dominant factors, and the hypothesis of higher heterozygosity. The first, so called hypothesis of dominance (Jones, 1917), is based on a frequently observed correlation between dominance and favorable factors, as a result of overlapping of dominant genes over recessive genes in homologous chromosomes. The second hypothesis, that is, the idea of overdominance, was originally proposed by G.H.Shull and E.M.East and definitely formulated by Hull (1945). According to this hypothesis, heterozygosis by itself promotes survival and development power.

From a genetic point of view it means that there are loci, the display of which proves to be more conspicuous in the heterozygous state than in the homozygous (Aa > AA > aa). In recent years most researchers admit the possibility of action of both mechanisms, while some of them (Lewis, 1955; Pontecorvo, 1955) believe that dominance and overdominance cannot be demarcated very strictly because of the presence of pseudoalleles and genes which are located very close to each other and exert a similar influence on development. Moreover, it should be taken into consideration that, in the process of evolution, the significance of each of the proposed mechanisms does not remain unchanged and, in the long run, both genetic mechanisms of heterosis can amalgamate into one.

The prolonged controversy between the exponents of both these hypothesis was successfully settled by Haldane (1955), who put forward a biochemical theory of heterosis. According to this theory, heterosis is conditioned by higher biochemical supply versatility of the hybrid zygote. The reason is that the heterozygote contains relative but different (and supplementary to each other) genetic products, or forms absolutely new materials. According to Kirpichnikov (1960a), such biochemical enrichment of hybrids is characteristic of heterosis under the conditions of any genetic mechanism which underlies it. The better biochemical maintenance of development leads to intensification of metabolic processes in hybrid specimens. The latter feature appears to be basic and most specific in the display of heterosis.

Hence, heterosis is related to dominance as well as overdominance, the effects of which may differ. This fact is of great importance, both for the most rational selection of groups for commercial hybridization and for appropriate organization of selection work on hybrids.

The relative effect of dominance and overdominance depends on the dimension and genetic structure of a population, on the intensity and direction of selection, and on many other factors. For example, heterosis observed in crossings of more or less inbred forms is a result of increasing the number of dominant genes and minimizing the adverse effect of recessives.

The value of heterosis in such cases will increase in proportion to the coefficient of inbreeding of parental forms and in proportion to the degree of the hybrid's heterozygosity. When inbreeding forms are crossed, one can observe either a rise in viability and general acceleration of growth or, more often, an intensification of some functions; in other words, the degree of luxuriance of hybrids is exhibited more distinctly. Such one-sided expression of heterosis would apparently be more profound if the crossing involves strains and breeds which for a long period were subject to purposeful selection.

Heterosis which results from crossing of specimens from highly heterogeneous outbreeding populations cannot be interpreted as a mere combination of actions of dominant factors. It is more probable that heterosis occurred in the crossing of different geographical populations. Natural species and sub-species are attributed to another mechanism, first of all to over-dominance. The over-dominant effect is inherent in few genes, but it always has great importance in the creation of heterosis. The value of heterosis, like dominance, is closely related to the level of heterozygosis.

Neverthless, the manifestation of hybrid advantage is more general. More often it is expressed in higher viability and general adjustability, increased resistance to diseases and to unfavorable effects of environment. An early disclosure of these properties is characteristic of hybrid specimens.

Heterosis in Fishes

Freshwater fishes have a number of biological peculiarities which facilitate the application of methods of commercial hybridization in pond fisheries. In addition, fishes manifest high fecundity, considerable magnitude of population, high level of natural heterozygosity, and explicit inbreeding depression in close breeding. The features determining higher heterozygosity of fish populations apparently condition the manifestation of heterosis in crossing of fishes. Taking these into account it appears possible to evolve a scheme of breeding best suited for obtaining and maintaining the maximum effect of heterosis under conditions of commercial hybridization.

In all probability, heterosis of hybrids of freshwater fishes is due to over-dominance. As experimental evidence of this, one can apparently consider demonstrated cases of heterosis on the molecular level in salmon hybrids, in whitefish hybrids (Kusakina, 1959, 1964) and in sunfishes (Manwell *et al.*, 1963). All authors revealed changes in the structure and properties of protein molecules of hybrid specimens. In salmon and whitefish the change was expressed in a non-specific rise of resistance of some hybrid proteins to the lethal effect of heat and alcohol, whereas in sunfishes it resulted in increased oxygen-combining properties of haemoglobin in the hybrids.

As indirect evidence of the significant role of the over-dominance mechanism, one can compare the cases of obtaining a considerable heterosis effect in inter-strain crosses with a low degree of inbreeding of initial lines, and the crossing of representatives of different species, sub-species and natural populations. As a rule, in moderately distant hybrids the heterosis effect shows a similarity, expressed in increased viability, higher rate of adaptability, and accelerated growth and development. The increased vitality is usually more conspicuous in the early stages of development. The results of numerous crossings in the sub-family Etheostomatinae is most illustrative of this (Hubbs, 1967). Out of 130 inter-specific and intra-specific crossings, 75 percent of the hybrids proved to be more viable at the moment of hatching or at the time feeding began.

Higher viability of eggs and young fry, as well as a reduced number of inheritable malformations, were found in hybrids of cultured carp x wild carp. It must be noted that these differences between hybrid and natural forms become more pronounced under unfavorable conditions of rearing (Kirpichnikov, 1959; Andriasheva, 1966). When raised in ponds the hybrids are often more viable than their parents. The hybrids withstand wintering and are more heat resistant. The same characteristic has been observed in whitefish.

Many hybrids are more resistant to parasitic diseases. Hybrids of salmon and sea trout, for example, are less susceptible to ichthyophthiriasis and bear it in a less severe form (Evropeitseva, 1963). Hybrid carp bred at the Ropsha fish rearing farm are more resistant to infectious gropsy (Kirpichnikov *et al.*, 1967).

Acceleration of growth and higher viability, which are characteristic of hybrids, are most distinct in the first months of rearing. The ability of hybrids to grow rapidly is especially conspicuous when conditions of nursing become unfavorable (abrupt variation in water temperature, over stocking or malnutrition). In such circumstances the rate of growth (relative weight increment) of hybrid specimens becomes higher than of the fastest growing parents, even in cases where the young hybrids initially do not exceed the parental forms (Lemanova, 1965). Such a general, non-specific nature of heterosis is mainly characteristic of hybrids produced by moderately distant hybridization and mostly in the case of intra-specific crossings (sub-specific, inter-population, race, inter-strain, inter-bred). In inter-specific hybrids, advantages are chiefly expressed in accelerated growth rather than in higher viability, although the combination of these features in hybrids is observed in some cases.

If the degree of taxonomic relationship is higher (when individuals of different genera are crossed) the effect of heterosis is hardly perceptible. The most general result of distant hybridization is a combination of valuable (from economic point of view) features of parental forms in hybrids without noticeable increase in viability and, as a rule, with intermediate growth rate. Only in rare occasions can inter-generic crosses give rise to heterosis in growth, due to disturbances in maturation and fecundity of the organism and related economy of its resources.

The dependence of results of htbridization on the degree of phylogenetic distance of crossing forms is reflected in many data on commercial hybridization. It must be mentioned, however, that because of non-compliance with methodological requirements in many experiments to determine the properties of hybrid forms. It is difficult to ascertain with adequate precision the presence and peculiarities of heterosis in different kinds of crossings.

Prospects of Commercial Fish Hybridization

Commercial hybridization implies crosses of species, subspecies, breeds and strains for the purpose of obtaining marketable hybrids of the first generation. These methods are now used with advantage in fish culture. In most cases, such crossings have yielded valuable hybrid forms, some of which are rather promising from the view point of economy for commercial rearing. They comprise both heterosis hybrids and hybrid forms with a favorable combination of parental features.

Intraspecific Crossings

Crosses of Cultured Carp with Wild Carp

Scattered Carp x Wild Carp (Kirpichnikov and Balkashina, 1935)

Explicit heterosis was advantageous, both in growth (age group) hybrids exceed cultured and wild carps by over 20-30 percent) and survival. Moreover, high resistance of hybrids to infectious dropsy is undoubted. By the second summer, growth heterosis disappears.

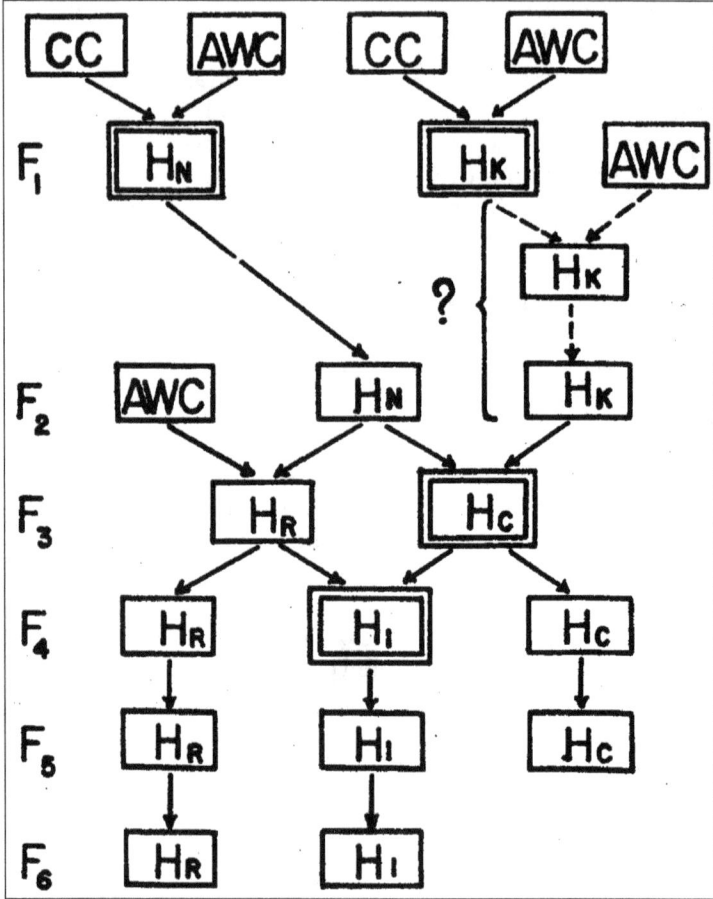

Chapter 12.1: Pedigree of Ropsha Carp.
CC: Cultured (Galician) mirror carp; AWC: Amur wild carp; H_N and H_K: Hybrid, carps of Novgorod and Kursk strains; H_R, H_I and H_{IR}: Hybrids of three selected strains (reversive, interlinear and intermediate). In double frames – heterosis combinations (H_C).

Galician Carp x Taparavan Carp (Saveljev, 1939)

Hybrids exceed the parental carp in weight by 37 percent, and under unfavorable rearing conditions the difference increases up to 110 percent. Mortality of hybrid fingerlings is 1.4 percent, whereas that of carp constitutes 24.8 percent. A large number of inherited malformations occur in Taparavan carp causing retardation in its growth and a high percentage of waste, especially in the wintering season.

Cultured Carp x Amur Wild Variety (Kirpichnikov, 1938, 1943, 1059; Andriasheva, 1966; Karpenko, 1966)

Heterosis in weight growth of hybrid young is 5-10 percent higher than that of initial forms and in viability heterosis expresses itself most clearly under unfavorable conditions of rearing (difference is in the range of 30-50 percent) and at early stages of development (hybrid larvae survive better than parents by 10-20 percent). Later, these differences become less and heterosis appears to fade away, although during the first or second years the hybrids retain larger sizes than wild carp or cultured carp.

Interstrain Crosses of Ropsha Carp

When crossed, the hybrids of the second generation of two different strains (Novgorod and Kursk) showed well defined heterosis in interstrain hybrids of the third generation. Such hybrids are noted for fast growth and high survival. In this respect they exceed not only the hybrids of the third generation obtained from pure-breeding of Novgorod and Kursk hybrids of the second generation, but also the hybrids of the first generation. In crossing hybrids of the third generation of two different branches, the hybrid advantage is still maintained, which, however, becomes poorly expressed in hybrids of the fourth generation.

Interbred Crosses of Carp

In the crossing of Ropsha carp with Ukrainian carp (Kuzema and Tomilenko, 1962) and in joint rearing of two-year-olds it was found that the interbred hybrid exceeds the Ukrainian carp in growth by 25 percent or more, while Ukrainian and Ropsha carp jointly reared in one pond did not differ from each other in their weight. Crossing Ropsha carp with indigenous Krasnodar carp of unknown origin, Kirpichnikov *et al.* (1967), found that the hybrids when reared to the same age as the parental types, weighed 3.49 g, being more than 1.5 times the weight of the indigenous ones (2.14 g), and also somewhat heavier than Ropsha carp (2.92 g). The mortality of fingerling of Ropsha carp was 33.6±1.2 percent, the mortality of the hybrids was 15.3±0.6 percent and of the indigenous carp 46.9±0.6 percent.

Interspecific Crossings

Ictalurus punctatus x *Ictalurus furcatus* (Channel Catfish x Blue Catfish) (Giudice, 1966)

Equal numbers of hybrids and parent species were marked and reared together. All fishes were of similar age and size when stocked. The hybrids gained 11 percent more weight than the channel catfish and 32 percent more than the blue catfish in the second summer. After three growing seasons they had gained 32 percent more than the channel catfish and 41 percent more than the blue catfish.

Ictiobus niger x *I. cyprinellus* (Giudice, 1964)

In this cross the hybrid forms exceed their parents by 30 percent in weight and 10 percent in length. The biggest hybrids were obtained by crossing *I. niger* x *I. bubalus*.

Esox niger x *E. lucius* (Chain Pickerel x Northern Pike) (Armbruster, 1966)

The weight of hybrids is intermediate, being 5.2 ounce in 72 days after hatching, whereas the northern pike weighed 8.5 ounce and the chain pickerel 3.7 ounce. The selectionist believes that negligible mortality of fingerlings can be attributed to the effect of hybrid power which did not influence the growth rate. Growth heterosis has been found also in crossing *Esox lucius* x *E. masquinongy* (Eddy, 1941; Black and Williamson, 1947).

Coregonus lavaretus x *C. albula ladogensis* (Lemanova, 1960; 1965)

Hybrids of cisco ripus x ludoga whitefish stand between the parents in size and weight, but are closer to ludoga which has a higher weight. The relative weight gain of hybrid young is considerably higher than that of the faster growing parent (ludoga), especially during the first two months of rearing in the nursery. Comparative experiments on survival of 0 group, carried out for three years, showed that these hybrids stocked with ludoga suffered far less mortality than the latter (survival was 87 percent and 64 percent respectively). Hybrid advantage in growth and survival appeared more explicit under unfavorable conditions of rearing (die-off and overheating) and during the winter season.

Coregonus lavaretus ludoga x *C. albula* (Ludoga x Cisco) (Gorbunova, 1962)

Both hybrid forms outgrew the parental forms. The fingerlings reared in ponds had the following weight : cisco, 3.3 g; ludoga, 8 g; ludoga x cisco, 22.3 g; cisco x ludoga, 10.8 g. Promising results were achieved in Latvia in experimental work on hybrids of *C. albula* x *C. lavaretus maraenoides* (Kokina, 1966). For acclimatization of white fish in the Ural area, hybrids of *C. albula ladogenesis* x *C. lavaretus maraenoides* were successfully stocked in natural water bodies (Pomerantsev and Nesterenko, 1960). The experiments with *Coregonus nasus* x *C. peled* have been taken up. The high growth rate and noticeable fatness of many coregonid hybrids are attributed by some researchers to the extended feeding range of these forms due to changes in the shape of the mouth and in the structure of pharyngeal apparatus. Coregonid hybrids make a better use of the food supply, which makes them preferable for transplantation and pond fisheries.

Crossing of *Tilapia* spp.

In many cases both interspecific and intraspecific hybridization reveal growth heterosis (Hickling, 1960). In crossing males of African strain of *Tilapia hornorum zanzibarica* (from Zanzibar) with females of Malayan line (from Malaysia) growth of hybrids was better than of the pure strains. It is worth noting that, in this cross, as in some others (*Tilapia mossambica* x *T. nilotica*), almost the whole progeny (up to 100 percent) are males. This trend of fast growing all-male hybrids is used successfully in pond fisheries because it permits control of population and prevents overcrowding of the water body.

The results of crossing males of *T. mossambica* (from Zanzibar) with females of two different species of *Tilapia* allowed Hickling (1960) to assume that Zanzibar fishes have a genetic mechanism of sex determination which differs from that of the Indonesian fishes.

Male-dominant hybrids (up to 75-100 percent) were found also in crosses of sunfishes *Lepomis gibosus* x *l. macrochirus*. Growth heterosis in hybrids of *Lepomis* was established long ago (Hubbs and Hubbs, 1933).

A clear expression of heterosis in growth and vitality is not observed in all crossings. In many cases hybrids are characterized by intermediate growth rate with a bias towards either parental form (mostly to maternal form). The survival rate of hybrids, as a rule, is not higher than that of the parental forms.

These circumstances, however, do not reduce the economic value of hybrids if they favorably combine valuable features of the parental forms and have adequate viability. Such a situation can frequently be observed in bi-generic crosses. Sturgeon hybrids illustrate this situation.

Bigeneric Crossings

Huso huso x *Acipenser ruthenus* (Nikoljukin and Timofeeva, 1954)

This hybrid represents a very valuable sturgeon which possesses parental features in their best combination, that is, fast growth rate of anadromous *H. huso* with the ability for early maturation of *A. ruthenus* as well as its ability to live in fresh water, ponds in particular. The remarkable fecundity of hybrids of the first generation allowed Nikoljukin to use these and other sturgeon hybrids (sturgeon x starlet; starlet x stellate sturgeon) for further selection work.

Catla catla x *Labeo rohita* – Crosses of Indian Carps (Chaudhuri, 1959)

The progeny inherits the small head of *L. rohita* and the large body of *C. catla*, and thus gains an advantage over parental forms in the weight of edible part. The F2 is characterized by good growth. The interspefic hybrid of Labeo (*L. rohita* x *L. calbasu*) grows fast as well.

Hypophthalmichthys molitrix x *Aristichthys nobilis* – Cross of Silver Carp with Big Head (Vinogradov and Erokhina, 1964; Tang, 1964)

These intergeneric hybrids show a good growth rate, especially under the conditions of lmited food supply in ponds. As in coregonid hybrids, this feature seems to be attributed to the intermediate structure of pharyngeal apparatus (Voropaev, 1968).

Cyprinus carpio x *Carassius carassius* (Nikoljukin, 1952)

Carp x crucian carp hybrids inherited the resistance of *C. carassius* and the high growth rate of common carp. The limited fecundity of hybrids made it possible for a long time to utilize only the first generation for the purpose of commercial fish farming. The fecundity was restored by application of a certain scheme of crossing of Ropsha hybrids with Chinese goldfish (Kuzema and Tomilenko, 1965).

This list of valuable hybrid forms, though far from complete, shows that commercial hybridization can be an effective means for increasing productivity of different fish species. The efficiency of hybridization can be increased considerably if the peculiarities of heterosis and other sequences of crossing are duly taken into account in the process of selecting particular groups for purposeful hybridization. If, for example, the main task of selectionist is to increase the general resistance of fish (which is frequently met with when water bodies in new climatic zones are to be stocked., it appears more efficient to use hybrids of moderately distant crossings (sub-species, inter-population, or interbred). The effect of heterosis in such cases is most probable. The favorable combination of valuable parental features in hybrids can also be achieved through distant crosses- inter-specific and even inter-generic.

Fixation of Heterosis

In many cases when hybrid forms are prolific, their utilization is not limited to raising the first generation for marketing. The obtaining of such hybrids can be considered as a beginning of selection work. Such synthetic crossing is followed by long selection activity directed at fixing desirable features in a breed and maintaining the heterosis effect in the following generations. This work cannot be successfully fulfilled without applying genetic methods of selection (based on the knowledge of the nature of heterosis) and taking into account specific biological and genetic features of hybrid and parental forms.

The method of maintaining heterosis in a chain of successive generations was worked out by Kirpichnikov (1960b) who has been engaged for many years in creating new breeds of northern (Ropsha) carp. Heterosis of hybrids of the first generation resulted from crossing European cultured carp with wild carp and were used by him as initial material for selection of the breed. The aim of selection was to obtain a fast growing breed of carp suitable for conditions in the northern area. *C. carpio haematopterus* noted for its high winter resistance qualities, was selected for crossing with cultured carp, even though effect of heterosis was found also in crosses of carp with the Volga wild carp and the Taparavan carp.

Specific features of cultured and wild carp, as objects of selection, urged the scientists to review a widely applied scheme of commercial hybridization implying creation of highly inbred lines and their consecutive crossing and selection of the best combinations. This program was formulated because of three interrelated factors;

1. The high level of natural heterozygosis in wild carp;
2. Noticeable inbreeding depression in close breeding of carp followed by considerable retardation in growth; and
3. Non-expedience of obtaining a high coefficient of inbreeding for carp because of slow maturation (In the north, in particular) and noticeable inbreeding depression.

It has been found that heterosis in carp can be rather considerable,

Even in crossings of specimens of relatively poor inbred strains. This means that even moderate inbreeding would lead to considerable differentiation of genetic systems and to genetic differences between strains. Their mutual crossing will be followed by increased heterozygosis and thus by heterosis.

If strains are subject to intensive selection, heterosis may be very strong, because selection, with special reference to growth and viability, will help in retention of heterozygosity and promote fixation and even increase hybrid power. Such a case of clearly manifested heterosis was observed in inter-strain hybrids of the third generation obtained from crossing of fishes of different strains. The heterosis effect continued to manifest in successive generations, although to a lesser degree.

Fixation of the heterosis effect was a result of applying a scheme of carp propagation which can evidently be recommended for experiments with a wide number of species similar to carp from the standpoint of genetic structure of the population. The basic principle of this method is a combination of selection with a system of two-line breeding accompanied by periodic crosses between them. Such a scheme makes it possible to maintain a high level of heterozygosis in a breed and to use heterosis in each generation, both in synthetic selection and selection without distant crossings.

The significance of line selection and moderate inbreeding followed by crossings in carp culture was underlined by Kuzema (1950) and Shaskolsky (1954). Line breeding systems have been employed in the process of creation and improvement of Ukrainian carp breeds (Kuzema, 1953).

Heterosis in relation to growth and resistance, which had been noticed in crossings of Ukrainian carp with Ropsha carp, indicates the possibility of using the heterosis effect in successive generations and in crossings of different carp breeds.

It must be noted that selectionists-ichthyologists do not yet pay proper attention to the problem of maintaining and increasing the heterosis effect in successive generations, though the efficiency of such a method is quite evident.

Another aspect should be also stressed. It appears difficult sometimes to keep at one fish farm the broodstock intended for conducting commercial crossings. This may be impeded by the biology of selected species, particularly by their ability to live and reproduce under certain climatic conditions. Moreover, in many cases (especially in inter-breeding) there is a danger of mixing parental stocks of initial forms with spawners of hybrid origin. In artificial breeding of hybrids one should not confine oneself to selection of groups for crossing with a view to rearing marketable hybrids of the first generation or to verify the possibility of using any scheme for fixing heterosis.

Summing Up

Industrial hybridization is and evidently will be of great importance in fish culture. It must be noted, however, that the experience accumulated in fish hybridization makes it necessary to be careful in recommending a method of commercial crossing. The main reason is the possible introduction of spawners of hybrid origin into the parental spawning stocks. In intra-specific crossings, mixing can easily occur due to a lack of phenotypic differences between initial groups. Thus, in many fish farms, some years after the beginning of work on hybrids of eastern carp, it was not possible to distinguish eastern carp from hybrids, and industrial hybridization lost its meaning.

Commercial crossings can be successful only in the case of well-controlled groups of parental stocks, such control can easily be exercised if fish farms receive ready spawners instead of young replacement stock. Fish farms should also be supplied with hybrid larvae reared at specialized selection farms.

In case of remote hybridization, uncontrolled crossings of highly specialized species and the introduction into the parental stocks of hybrid fish can lead to loss of valuable properties of parents. It is especially dangerous when fertile hybrids have the opportunity (and even allowed) to go into natural water bodies. Before new hybrid forms are recommended for commercial propagation, it is necessary to carry out accurate experiments which will allow the evaluation of economic properties of new hybrids and to determine their advantage over both parent groups. Such work must be carried out on a large scale with multiple replication and repetition of hatching and rearing to permit statistical appraisal of the differences between hybrid and parental forms.

The basic requirements for conducting experiments intended for verifying haterosis should be maximum equalization of conditions for hatching eggs and rearing young and comparable stocking rates.

The establishment of a permanent regime of hatching (water temperature, content of oxygen and metabolic products in water), a precise count of the initial number of eggs, and measurements of levels of waste in the process of embryonic and post-hatching development are of great importance. The stocking and rearing requires equalization of initial weight of groups to be compared, otherwise the results of rearing would not be representative. It is desirable to apply both separate and joint schemes of rearing of all comparable combinations. Separate rearing should be effected in serial ponds with no less than three replications. Where there is lack of such ponds (which is the most important obstacle in hybridization work), pond rearing can be replaced by rearing in aquaria. Both parental forms should not act as test variants (controls) in all such experiments.

It has to be noted, however, that in some cases; mainly in inter-generic hybridization, the comparison of hybrids with both parental forms appears difficult due to considerable differences in growth rate because of specific peculiarities etc.

It is absolutely necessary to distinguish between the meanings of heterosis and hybrid. In any individual case of hybridization it is necessary to make a very thorough comparison of hybrid and parental forms to ascertain the hybrid's advantages. The compliance of methodological requirements in such comparisons is of utmost significance because it makes it possible not only to detect certain features of hybrid and assess their properties in relation to economy, but it promotes also a deeper study of problems of heterosis as a whole.

All this indicates that industrial hybridization should be conducted only as specialized, selective genetic work. It should be ensured that there is compliance with the requirements of preliminary assessment of economic value of hybrids with strict control of the state of parental stocks and maintenance of purity of initial forms in natural water bodies..

Chapter 13

Practical Significance of Hybridization of Sturgeons in U.S.S.R.

Natural Hybridization of Sturgeons

Sturgeons are known to interbreed under natural conditions, giving rise to viable and sometimes fertile inter-specific and inter-generic hybrids. Hybrids have been described from crosses of various combinations of almost all species of the family, Acipenseridae. L.S. Berg in his book, " Freshwater fishes of the USSR and the neighboring countries" (1948) described the hybrid forms of this family from the following crossings :

1. Kaluga (*Huso dauricus*) x Amur sturgeon (*Acipenser schrencki*)
2. Beluga (*Huso huso*) x Spiny sturgeon (*A. nudiventris*)
3. Beluga (*H. huso*) x Sturgeon (*A. guldenstadti*)
4. Beluga (*H. huso*) x Stellate sturgeon (*A. stellatus*)
5. Spiny sturgeon (*A. nudiventris*) x Stellate sturgeon (*A. stellatus*)
6. Sterlet (*A. ruthenus*) x Sturgeon (*A. guldenstadti*)
7. Sterlet (*A. ruthenus*) x Stellate sturgeon (*A. stellatus*)
8. Sturgeon (*A. guldenstadti*) x Stellate sturgeon (*A. stellatus*)
9. Siberian sturgeon (*A. baeri*) x Sterlet (*A. ruthenus*)

This list does not include the whole variety of acipenserid hybrids that can occur in nature, as for instance the hybrid between beluga and starlet. The capacity of Acipenseridae to interbreed resulted in higher variability and formation of different varieties.

Artificial Hybridization

The first artificial hybridization of Acipenseridae was accomplished by F.V. Ovsyannikov, who as early as 1869 fertilized the eggs of starlet (*A. ruthenus*) by the sperm of sturgeon (*A. guldenstadti*) and stellate sturgeon (*A. stellatus*), and drew attention to the importance of research in this direction. Later, however, experiments on hybridization of Acipenseridae were discontinued for a long time. Investigations were resumed in 1949 in the Volga River, in line with the research program of VNIRO (All-Union Research Institute of Marine Fisheries and Oceanography). The task set was to combine in the hybrid organisms the properties of such large anadromus Acipenseridae as the beluga (*H. huso*) and the sturgeon (*A. guldenstadti*) with the properties of the freshwater starlet. It was assumed that products of the cross between starlet and sturgeon or beluga, with increased adaptive properties and heterosis, would be distinguished for their economic value as compared to the parental species. Since hybridization involved the starlet, which attain sexual maturity earlier than other Acipenseridae, it was assumed that the hybrids produced would differ favorably in this respect from late-maturing anadromous Acipenseridae. Experimental research confirmed the soundness of these suppositions. It is expedient to make use of the heterosis of the hybrid forms of Acipenseridae when rearing them in freshwater ponds and reservoirs.

Pond Breeding of Hybrids

Though Acipenseridae are suited for commercial breeding in ponds they have seldom been used for this purpose in the past. Sterlet were successfully bred in ponds for many years, but this practice has now been nearly abandoned because of the slow rate of growth of this fish.

Of interest in this respect are some hybrid forms which adapt themselves more easily to pond conditions that are unusual and hardly suitable for Acipenseridae. Besides, the favorable characteristics of hybrids based on heterosis, namely, rapid growth, greater viability, and early sexual maturity may be advantageously used.

Among the acipenserid hybrids studied, the most promising in respect to fisheries is the hybrid between beluga and starlet, which favorably combines the valuable properties of both parental species; the rapid growth rate beluga and the early maturity of starlet.

Growth

Even in ponds of low fish productivity, where the potential speed of growth of the hybrid could not manifest itself to the full, it reached the weight of 500 g by the end of the second summer, while starlet under the same conditions grew much slower. In one of the ponds in Saratov region which has greater food reserves, the hybrid fingerlings reached the mean weight of 74 g in one summer, the yield of fish being 580 kg/ha. In the other ponds where the hybrids were cultured along with carp, the fish production did not exceed 300 kg/ha, the growth of the hybrid being slower as they could not compete successfully with carp for benthic food.

The potential rate of growth of hybrids is quite high when sufficient food is available, one summer fish growing to an average weight of 0.5 kg. this potential for rapid growth can be used advantageously by rearing them in ponds to marketable size with the help of intensive artificial feeding. Within two growing seasons it is possible to produce marketable hybrids weighing nearly 1 kg.

Food

The favorite food of very young hybrids is larvae of Chironomidae, called bloodworms. Later, those that have inherited predatory instincts from beluga start feeding on larger animals, such as tadpoles, small frogs and any small inactive fish. Fingerlings of hybrids can attain average weight of 90-100 g in ponds with abundant food reserve, whereas in the sea, where the standing crop of benthos is many times higher than in ponds, they average 500 g. Besides using benthic invertebrates, hybrids in the sea feed on numerous slow-swimming bottom fish, mainly Gobiidae, which are not to be found in ponds.

To supply Acipenseridae with a sufficient quantity of food in ponds, supplementary artificial feeds should be introduced on feed tray. Since all Acipenseridae in nature feed only on animal food they should be fed on food of animal origin too; scrap fish or fish of little value, wastes of fish and meat processing plants, various invertebrates, such as, Gammaridae etc. It is desirable that all these products be given fresh or frozen, and if this is impossible, these feeds can be preserved with sodium pyrosulphate. The conversion ratio (food per unit of weight gain of fish) may be assumed as equal to 7.

Stocking Rate

The stocking rate of young may vary, approximately 15000 to 20000 individuals per hectare, depending on the natural fish productivity of the pond and on the possibilities of regularly introducing supplementary food, starting, at least, from the second half of the growing period.

The stocking rate of yearlings is 2000 individuals per ha and two-year-old hybrids are fed intensively during the whole growing period. If the rate of stocking is higher feeding should assume still more importance, since the capacity for natural fish productivity of pond decreases. The rearing practice should then acquire the character of tank fish culture.

Yield

To raise fish production in ponds, it is advisable to rear hybrids together with herbivorous fishes of the family Cyprinidae, provided they are of the same age group. The stocking rates of hybrids, mentioned may be retained, but it is desirable to lower the usual stocking rate of herbivorous fishes by approximately 30 percent.

In the autumn of the second growing season, two-year-olds weighing not less than 800 g are delivered to the market. A number of hybrids that have not reached marketable size are left to grow in the pond for a third summer.

Due to increased viability, hybrids winter well in the usual carp ponds; the mortality rate, even among one summer fish, being quite insignificant.

The results of rearing hybrids in the Rostov region are of interest. In 1965, a pond with area of 0.1 ha, was stocked with 3700 young hybrids weighing 4 g each. The natural food resources were extremely poor; the standing crop of benthos averaged 0.13 g per square meter. The young were fed mainly on minced fish. The yield (survival) of young of the year, with average weight of 62 g was 58.4 percent of the number of fingerlings stocked. Fish productivity equaled to 1330 kg per hectare. Two-year-old hybrids averaged 0.9 kg the maximum individual weight being 1.8 kg.

In 1967 an experiment was undertaken in culturing mixed age groups of one-summer fish and three-year-olds. Pond of 0.1 ha area was stocked with 1000 fry of 4 g mean weight and 213 two-year olds with mean weight of 700 g, the total stocking rate being 12130 individuals per ha. As a result, fingerlings of average weight of 85.2 g (yield-72 percent), and three-year-olds of average weight of 1.94 kg (yield-80 percent) were produced. Fish productivity of the pond was 570 kg/ha for fingerlings and 2200 kg/ha for three-year-olds, the total productivity being 2770 kg/ha.

Reproduction

Of specially great significance is the unimpaired ability of the hybrid of beluga and starlet (different genera) to reproduce. This is a very rare phenomenon in inter-generic hybrids. The onset of sexual maturity in the hybrid is early; in males- at the age of 3-4 years; in females- from the age of 6-7 years. The size of eggs of the hybrid is nearly the same as that of the stellate, but smaller than *Huso huso*.

The fecundity of the hybrid and the high flavor qualities both of the flesh and of the caviar increase its value in commercial culture. The possibility of producing second and subsequent hybrid generations opens prospects of selective breeding of new varieties of Acipenseridae.

Due to heterosis and the heightened adaptive plasticity, the sexual products in the hybrid broodstock mature even when they are bred and reared in ponds throughout their whole life cycle, never entering a river or a sea. The fish cultural problem of producing progeny of Acipenseridae has thus been solved, even raising them entirely under pond conditions. Here the positive role evidently has been played by the method of distant hybridization. However, numerous attempts to achieve this with thorough bred Acipenseridae have not been successful so far.

Viability

The high viability and adaptability of the hybrid is illustrated by a batch of fish that for a long time (about 6 years) starved in the ponds to such a degree that they not only did not gain weight during the growing season, but many of them lost weight, could in the subsequent years when fed well produce both sperm and ripe eggs.

The unusual viability of the hybrid made it possible to use the method of stripping eggs, by incision, from live females that had received hypophyseal injection, after which the incised abdominal wall was sutured. The operation was well endured by the hybrid females. By this technique, I.A. Burtsev (in 1967) obtained eggs for fertilization and production of the second hybrid progeny from five females averaging 8.3 kg. The eggs from one female weighed one kg. The females which were operated on were returned to the pond of the Aksay fish hatchery. Examination of the females in October, when transplanting them to wintering pond, showed that they were in good condition; the sutures were completely grown over and hardly noticeable as pale stripes. The females not only survived but gained considerable weight during the growing season, from 1.7 to 3.6 kg. Thus the operation did not have any adverse effect on the subsequent condition and growth of the females. The technique of successfully obtaining eggs from living females, applied for the first time in sturgeon culture, might enable the repeated use of individual brood fish. This makes it possible to select females by the quality of their offspring, which must play a significant role in selection work.

Hybrid F2

The second brood of hybrids of beluga and starlet (F2) was first obtained in the sturgeon hatchery, located in the Don River delta. The first was produced in 1966, and a considerably greater number (33000 fry, reared to average weight of 3.5 g) was produced in 1967. The fish production indices (percentage of fertilized eggs, survival of young in reservoirs and ponds) proved in the majority of the cases to be not lower, but at times even higher, than in the hybrid F1.

Nearly all the young F2 offspring were transported to the Don fish rearing farm for pond culture of the young-of-the-year and for carrying out selection experiments. Some of the young produced from the eggs of each of the five females were introduced separately into five small ponds, the one-summer fish averaging 79.4 g, the maximum individual weight being 250 g.

In rearing one-summer fish of hybrid F2 and silver carp together, the total fish productivity of one of the ponds equaled 1300 kg/ha.

Morphologically, both hybrid F1 and F2 take an intermediate position between the parental species, but in fact there is a much greater variety in F2 individuals. As compared to F1, it especially varies in body colouration. The individual variability in F2, of some countable traits and measurements, is also higher.

It is of interest to note that cytogenetically both broods resemble each other and the parental species: the modal number of chromosomes in all of them is 60, varying somewhat greater in F2 than in F1 and the parental species. This follows from the comparison of the variation sets. Consequently, the hybrid is characterized by a heightened variability both morphologically and cytogenetically (Nikoljukin, 1966).

The possibility of producing new breeds of the beluga x starlet hybrid should find application in the production of new breeds with stable heredity, especially the pond breed of Acipenseridae. On the other hand, the possibility of producing progeny from hybrid broodstock reared under pond conditions opens prospects of organizing specialized sturgeon farm ponds covering all the processes of fish culture; from obtaining fertilized eggs to producing a marketable crop without obtaining hybrid fry from sturgeon hatcheries.

Due to the wide range of adaptability of the beluga x starlet hybrid, it can dwell in purely freshwater ponds, or in brackish water bodies or in salt water seas.

In the fifth year of life many individual hybrid attained the weight of 8 kg and mature males with ripe sperm were found.

The basis for obtaining this new breed (the hybrid between beluga and starlet), the young of which have spread in the Bay and Sea proper, where they grow rapidly, predating on other fish. The survival rate of the hybrid is high. In the autumn of 1964 young hybrids prevailed in the commercial catches constituting 50 percent of the total catch of young Acipenseridae.

The young of the year feed mainly on Musidacea and fish, whereas two-year-old hybrids feed exclusively on fish (Gobiidae and *Clupeonella* sp).

Intensive feeding accelerates the sexual maturity of the hybrids. The males become mature by about 3 years of age. In the fifth year of life, in 1967, a more intensive spawning migration of males from the sea into the river was observed. By this time the hybrids had attained the weight of 13 kg.

Chapter 14

Cytogenetics in Fish Breeding

Chromosomes were first described by Strasburger (Germany) in 1875. In 1882, Flemming (Germany) experimentally proved that during cell division, chromosomes split longitudinally. In 1883, Van Benedin (Belgium) discovered that when gamets are formed, the number of chromosome is reduced to half of that present in the body cells. In 1884-85, Hertwig, Strasburger, Kollikar, Weismann and Flemming independently contributed to the knowledge of cell division. In 1885, Rabi (Austria) proved the individuality of chromosomes.

Chromosomes studies in fish started as early as 1891 when Bohn investigated the chromosome number of *Salmo trutta fario*. This study was based on the histologically sectioned gonadal material, particularly the testicular tissue. It was followed by the use of hypotonic treatment (Sharma, 1960), colchicines (Roberts, 1964) and the flame drying method (Denion and Howell, 1969) in fish chromosome handling.

In fact, the methodologies of human and mammalian chromosome preparation were applied to fish tissues, such as, gills, kidney and other soft organs. Since then several cytogenetic studies have been carried out with fish.

Chromosomes

In all eukaryotic organisms including fishes, the DNA molecules in the nucleus are combined into proteins, mainly histones, to make chromosomes. The chromosomes are darkly stained small bodies present in the nucleus and determine the mechanism of inheritance. In 1902, W.S. Sutton and T. Boveri suggested that chromosomes were the physical structures which act as messengers of heredity. After specific staining thay can be clearly seen during mitotic cell division, particularly during the metaphase. Both the size and shape of chromosomes differ in different species of fish. The chromosome number in the cells of the body is normally made up of two sets, one of maternal and one of paternal origin, and is termed as diploid.

A single set of chromosomes is known as a haploid set. When more than two sets of chromosomes occur in cells, the condition is termed polyploidy.

The material of chromosomes is called chromatin. It contains euchromatin which stains but lightly and heterochromatin which stains darkly. This differential staining is the basic for most of the methods now used for detecting details of the structure of chromosomes. Euchromatin is believed to contain the genes in a linear array like beads on a string, whilst heterochromatin is regarded as genetically inert and to have a function in maintaining the structural integrity of the chromosome and perhaps regulating gene expression. Heterochromatin is mostly made up of highly repeated simple sequences of DNA. Two other features of chromosomes are the ends, called telomers, and a constriction called the centromere. The telomeres are a stable entity, usually heterochromatic and are for maintaining the integrity of chromosomes threads

Fish chromosomal complements contained chromosomes of two, three or even four types. In a number of teleostean fishes additionally very small "microchromosomes" had been found along with the larger ones and it is very difficult to quantify such chromosomes. The role of microchromosomes is not fully understood although some authors suggest that they contain the "redundant" genetic material necessary for the cells where there is increased protein synthesis. In salmonids, "satellite chromosomes" have been reported. In these chromosomes small region are separated from the main body by narrow strangulations.

Each chromosome contains two identical parallel structures of chromatids. Each chromatid in turn consists of one or more thin filaments called chromonemata containing characteristic condensed stainable regions called chromomeres. Chromonema contains alternating thick and thin regions. Centromere or kinetochore divides the chromosomes into two parts. The position of centromere varies from chromosome to chromosome. Centromere forms the site of implantation of the microtubules that constitute chromosomal spindle fibers. The chromosomal ends are known as telomeres. Each extremity of the chromosome has a polarity and therefore it prevents other chromosomal segments to be fused with it.

Morphologically a chromosome in mitotic metaphase has only two distinguishing features, its length and a transverse constriction that makes the position of the centromere. From the chromosome length and centromere position three factors can be calculated. The Centromeric Index, The Arm Ratio and The Relative Length of the chromosome. The first two factors, Centromere Index and Arm Ratio, tell us all about the chromosome itself. The relative length tells us about the size of the chromosome in relation to the other chromosomes in the set.

1. **Centromeric Index:** It is defined as the length of the shorter of the two chromosome arms multiplied by 100 and divided by the length of the whole chromosome and expressed as percentage.

2. **Arm Ratio**: It is defined as the length of the longer arm of the chromosome divided by the length of the shorter arm. It is always greater than 1.

3. **Relative Length**: It is defined as the length of the whole chromosome multiplied by 100 and divided by the total length of all the chromosomes

in the haploid set including one being measured and expressed as a percentage.

Genes, the elementary units of heredity, are located throughout the length of the chromosome. In fishes and other vertebrate animals, each chromosome is known to contain hundreds and perhaps even thousands of genes. The basic of the chromonema is represented by DNA or deoxyribonucleic acid. The structure of DNA was elucidated by Watson and Crick (1953).

Types of Chromosomes

Fish chromosomes possess the same morphology as is witnessed in other species of animals. Crromosoms are classified as follows, according to the position of centromere.

1. **Metacentric:** Centromere is located approximately in the center of the chromosome.
2. **Sub-metacentric :** Centromere is located midway between the center and end of the chromosome.
3. **Sub-telocentric:** Centromere is located at sub-terminal position of the chromosome.
4. **Acrocentric or telocentric:** Centromere is located at the end (terminal) of the chrosome.

On the basis of arm ratio (which is calculated by dividing the long arm of a chromosome by the short arm), the following types of chromosomes can be categorized.

1.00-1.70

     Metacentric chromosome

1.71-3.00

     Sub-metacentric chromosome

3.01-7.00

     Sub-telocentric chromosome

7.01   and above

Acrocentric/Telocentric

In size, however, the fish chromosomes are much smaller than the chromosomes of other groups like mammals and amphibians. But some fish groups (namely, salmonids, dipnoans) have chromosomes as large as or even larger than those of animals. The karyotype of sturgeons show a full range of size of chromosomes extending from micro to macro chromosomes.

Regarding the number of chromosomes in fishes, no other group shows such a wide range as do the fishes. Their numbers range from 2n=12 in *Gonostoma bathyphylum* or 2n=16 in *Spheririchthys psphromonoides* to 2n=446 in *Diptychus dipogen*. Chromosome number is constant for each species. Number has no specific significance and does not indicate in any way the evolutionary advancement of a species.

Sex Chromosomes

Sex chromosomes are those chromosomes that usually determine an individual's sex and the pairs of sex chromosomes are often morphologically different in the two sexes. Scientists studying fish chromosomes were unable to find heterochromatin in them for quite a long time. True heterochromosomes, differing in their size and structure were found recently in a number of species. In a review of the karyotypes of 1700 bony fishes, heteromorphic sex chromosomes have been found in 104 species. Almost all types of sex chromosomes indicating male heterogamety multiple sex elements have been reported in fishes. Male heterogamety of the type XX female, XY male was the most common and ZZ male, ZW female were less common.

Autosomes

Autosomes are the other pairs of chromosomes and are morphologically the same in males and females.

There are many protocols for preparing a karyotype from peripheral blood lymphocytes. In this connection, a rather standard series of steps are involved.

A sample of blood is drawn and coagulation is prevented by addition of heparin.

Mononuclear cells are purified from the blood by centrifugation through a dense medium that allows red cells and granulocytes to pellet, but retards the mononuclear cells (lymphocytes and monocytes).

The mononuclear cells are cultured for 3-4 days in the presence of a mitogen like phytohaemagglutinin, which stimulates the lymphocytes to proliferate madly.

At the end of the culture period when there is a large population of dividing cells, the culture is treated with a drug such as colcemid. Which disrupts mitotic spindles and prevents completion of mitosis. This greatly enriches the population of metaphase cells.

The lymphocytes are harvested and treated briefly with a hypotonic solution. This makes the nuclei swell osmotically and this greatly aids in getting preparations in which the chromosomes do not lie on top of one another.

The swollen cells are fixed, dropped on to a microscope slide and dried.

Slides are stained after treatment to induce a banding pattern.

Once stained slides are prepared, they are scanned to identify good chromosome spreads (that is, the chromosomes are not too long or too compact and are not overlapping), which are photographed. The photos then are cut out in an orderly manner limited to grouping like-sized chromosomes together in pairs and if the chromosomes are banded.

Applicability of Fish Cytogenetics

1. The most important precondition for the applied use of fish cytogenetics is to establish the "Standard Karyotype" for every species of commercial importance. Among fishes, standard karyotypes are still wanting except for 2 or 3 species. Therefore, there is an urgent need to establish standard karyotype by the use of strict morphometric criteria, banding analysis and computer programming.

2. On the basis of standard karyotypes, efforts should be made to distinguish the abnormal karyotypes and establish their link with the affected traits.

3. Since fishes are confined to closed environment, these are directly affected by any pollutant in the water much more persistently. Therefore, fishes can form an excellent material for monitoring genotoxic effects of the various harmful pollutants being drained into the aquatic ecosystem.

4. Sex chromosomes may be studied in important fish species with more authentic methods like banding and the use of sex specific probes, as is done in humans and domestic animals. The determination of sex chromosome mechanism forms very basic information needed for application in fish reproductive biology.

5. Another application of cytogenetics is related to the manipulation of chromosome sets in fishes. The manipulation chromosome set includes induced gynogenesis and androgenesis, induced polyploidy, sex reversal, inter-specific and intra-specific hybridization and production of transgenic fish. Cytogenetics methods clearly find a use in all manipulations.

6. Intra-specific hybridization has played a pivotal role in plant genetics and crop productivity. Unlike inter-specific hybrids, the intra-specific hybrids are fertile. Cytogenetic races of various culturable fin and shell fish species can be made use of in experimental hybridization.

The Molecular Marker for Sex Determination and Breeding

The advances in molecular biology techniques have generated a number of DNA markers for the identification of genetic polymorphism. Random Amplified Polymorphic DNA (RAPD) method based on the polymerase chain reaction (PCR) has commonly used techniques for developing DNA markers.

The molecular marker techniques have been introduced to be an effective tool for both the identification of sex-specific genetic markers and of sex control in fish. With this aim sex-specific DNA markers are used for hatchery management. Sex identification for the fish at early ages can reduce broodstock rearing costs. RAPD markers were successfully used in discrimination of sexes in Nile tilapia fish (*Oreochromis niloticus*) using linear discriminant function analysis. The results provide support for the view that major genetical sex determining factors exist in tilapia.

Detection of Hybrid Introgression

In species for which diagnostic markers have been developed, these markers can be readily used for detection of hybridization in cultured or wild stocks. This can have significant consequences if it occurs in hatchery stocks. Padhi and Mandal (1997) used RFLP to identify F1 hybrids between catla, rohu and mrigal in fry from a mixed species spawning pool in a West Bengal hatchery indicating that "incidental" hybridization can occur in such systems making it likely that hybrids would eventually enter the brood stock. Allozymes were the tools of choice by Simsen *et al.* (2005) in their identification of hybridization in between Chinese carp species in Bangladesh, with mitochondrial markers used to identify how such hybrids might have been produced. These findings were confirmed to some extent by Mia *et al.* (2005) in their investigations into the extent of uncontrolled hybridization in Bangladesh carp hatcheries using microsatellite markers.

Parentage Assignment

Because of their high polymorphic nature and the ability to combine (through multiplexing) several markers in a single reaction, microsatellites are generally the markers of choice for parentage assignment. These enable progeny to be identified to parental pairings after they have been reared in communal groups. One application of this technique is to determine the consequences of spawning or mating designs on genetic contributions and effective population sizes. It has been shown that relatively small proportions of males contribute to progeny in pooled matings in Barramundi *Lates calcarifer* and tilapia leading to low effective population sizes.

Parentage assignment can be particularly applied usefully in enhancing selective breeding programs based on family designs which often require families to be reared separately for several months until they are big enough to be tagged. This may then increase the probability of confounding environmental effects on the trait in question. Parentage assignment allows progeny to be pooled at a very early stage (preferably at the same age to minimize interaction effects between families) and to be identified to family at harvest. One of the drawbacks of this method is that contributions from different families can vary substantially due to differential larval survival and size-based grading, especially when pooled very early and this may result in numbers being highly skewed to just a few families, reducing effective population sizes. One way to circumvent this problem, even in mass selection, is to use percentage assignment to identify selected progeny and then "walk back" from those with the highest breeding value for the traits being selected to those with lower values until enough pairs have been selected from enough families to meet the targeted effective population size (Sonnesen, 2005).

Chapter 15

Gene Transfer Technology: Production of Transgenic Fish

The main aim of transgenesis is to achieve specific traits in an organism by the introduction of a foreign gene. It is defined as, "the introduction of exogenous DNA into the genome such that it is stably maintained in a heritable manner (Khoo, 1995; Choudhary *et al.*, 2005). Transgenesis or gene transfer is the process of *in vitro* transfer of exogenous and usually heterologous genes or recombinant gene constructs into the genome of organisms. Transgenic animals are considered under the category of genetically modified organisms (GMOs) on which actual gene manipulation has been performed. The International Council for the Exploration of the Sea (ICES) defines a GMO as "an organism in which the genetic material has been altered anthropogenically by means of gene or cell technologies".Such technologies include, isolation, characterization and modification of genes and their introduction into living cells or viruses of DNA as well as techniques for the production involving cells with new combinations of genetic material by the fusion of two or more cells. The European defines a GMO as "an organism in which the genetic materialhas been altered in a way that does not occur naturally by mating and/or natural recombination (Bartley, 1999). Generally, GMO are organisms in which genetic material has been purposely altered through genetic engineering in a way that does not occur naturally. It is obvious that the definition of GMO varies considerably and the scope can vary from one user to another.

Production of Transgenic Fish

A transgene used in producing transgenic fish, is a recombinant gene construct that will produce a gene product at appropriate levels in the desired tissue(s) at the desired time(s). The prototype of a transgene is usually constructed in a plasmid to contain an appropriate promoter/enhancer element and the structural gene (Chen *et al.*, 1995).

Depending on the purpose of the gene transfer studies, transgenes can be grouped into three main types; (i) gain-of-function, (ii) reporter function and (iii) loss-of-function.

The gain-of-function transgenes are designed to add new functions to the transgenic individuals or to facilitate the identification of the transgenic individuals if the genes are expressed properly in the transgenic individuals. Transgenes containing the structural genes of mammalian and fish growth hormones (GH or their cDNAs) fused to functional promoters such as chicken and fish (3-actin gene

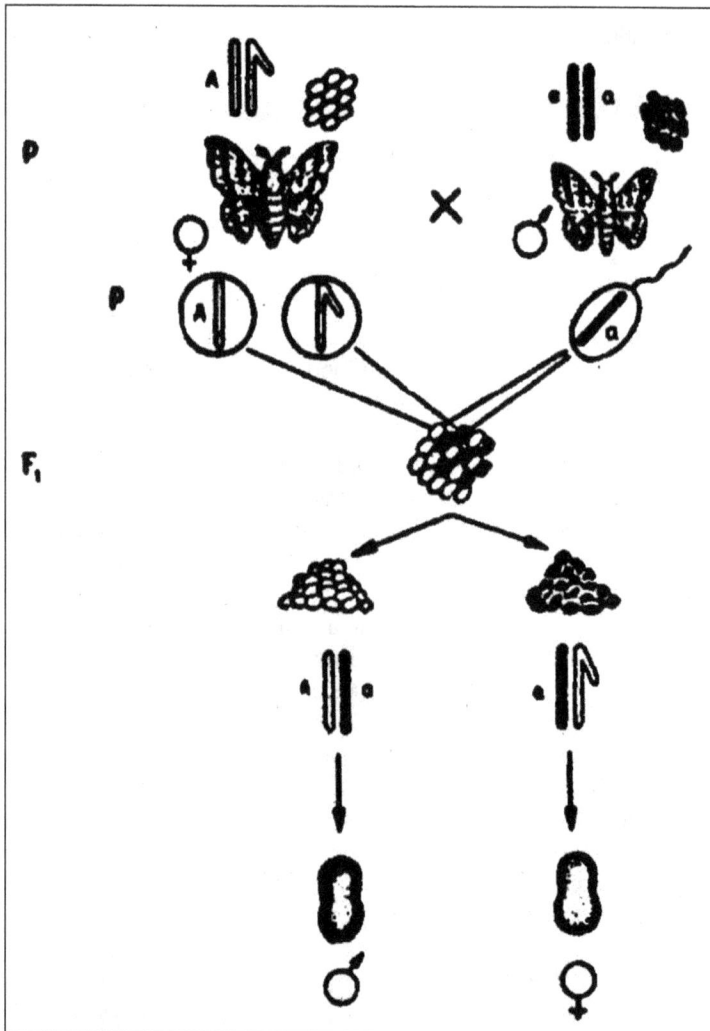

Figure 15.1: Scheme for Inheritance of Sex-Linked Character of Silkworm (Bombyx) Egg Colouring. A-factor of light colouring of silkworrm eggs, a-factor of dark colouring.

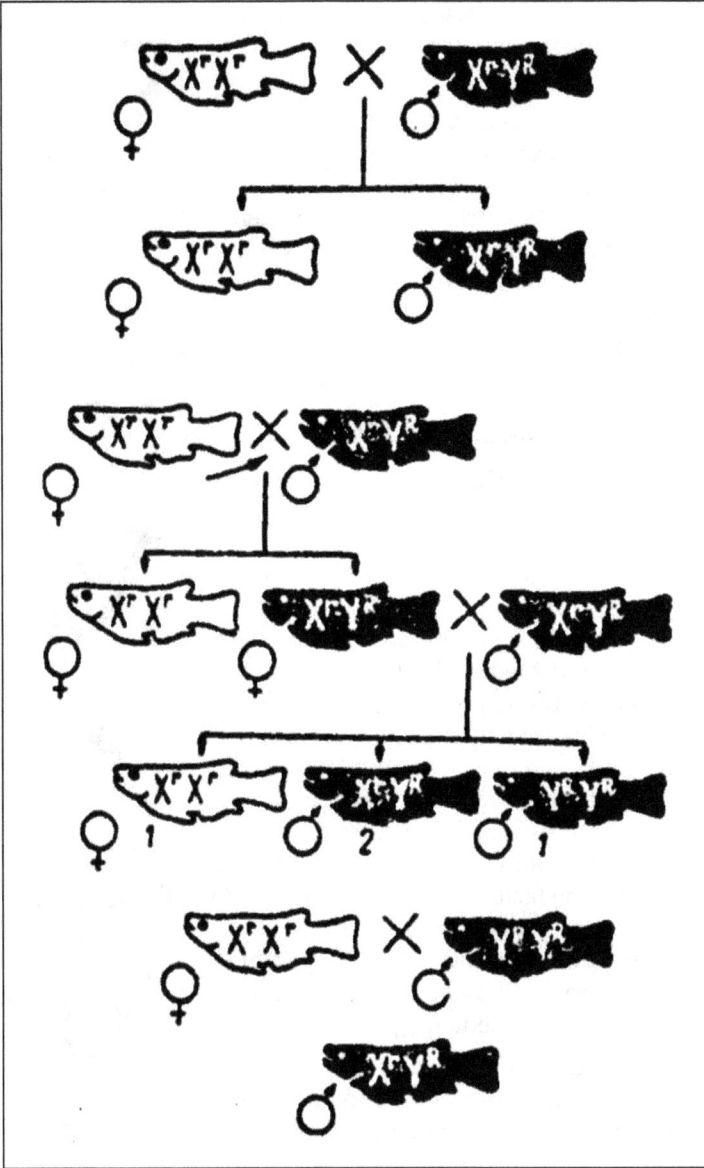

Figure 15.2: Phenotypical Sex Determination in Fish.

promoters) are examples of the gain-of-function transgene constructs. Expression of the GH transgenes in growth enhancement (Zhang, *et.al*, 1990; DU *et al.*, 1992; Chen *et al.*, 1993).

The reporter genes are commonly used to identify the success of gene transfer effort. A more important function of a reporter gene is used to identify and measure the strength of a promoter/enhancer element. For example, bacterial chloramphenicol acetyl transferase (CAT), (3-galactosidase, or luciferase genes fused to functional promoters are one type of the reporter function transgenes.

The loss-of-function transgenes are constructed for interfering with the expression of host genes. Such genes might encode an antisense RNA to interfere with the processing and transcription of normal genes or catalytic RNA (a ribozyme) that can cleave specific mRNAs and thereby cancel the production of the normal gene product (Cotton and Jennings, 1989). These genes can be used to produce disease-resistant transgenic broodstocks for aquaculture.

Gene Transfer Technique in Fish

The foundation of gene transfer research was actually laid as early as 1910, when embryologists experimented with injecting cellular material into frog eggs (Gurdon and Melton, 1981). By the early 1970s, it was apparent that gene transfer technology could provide great insight into the function of DNA sequences (Gurdon and Melton, 1981). Technological advances in the isolation and proliferation of eukaryotic genes, coupled with the development of microinjection procedures for amphibian eggs, resulted in the rapid expansion of gene transfer research. At present the methods available for transfer of the gene into eukaryotic cells may be classified into two categories; facilitated and direct delivery. The facilitated techniques include electroporation, transfection of DNA, lipofection, virus-tagged gene delivery and receptor-mediated targeted gene delivery. The direct method includes microinjection and particle-mediated gene transfer.

Recently, the use of embryonic stem cells (ESC) as a method for inducing transgenesis has been advocated. These cells (blastomeres) are undifferentiated and remain totipotent/pluripotent, so they can be manipulated *in vitro* and subsequently reintroduced into early embryos where thay can contribute to the germ line of the host (Melamed *et al.*, 2002). Similarly, premprdial germ cell (PGC) transplantation technique has also been developed for fish, providing another useful approach for the production of transgenic fish (Devlin, 2004).

The microinjection method is suitable for relatively small numbers of organisms whereas other methods are

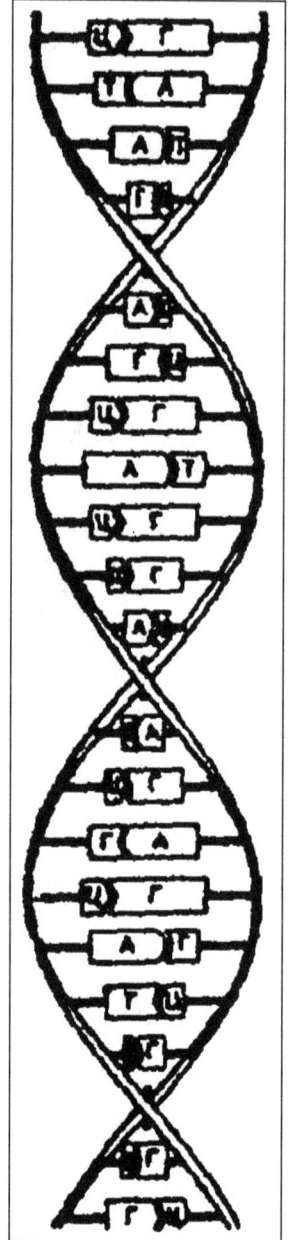

Figure 15.3: Structure Model of DNA according to Watson and Crick. Combination of two polynucleotide chains by hydrogen bonds, forming the so-called steps of a rope ladder.

more suitable for mass gene transfer for the production of transgenic individuals. Moreover, mass gene transfer systems are especially beneficial for recombinant DNA research in fish and shell fish because of the high fecundity and high embryo mortality of some of the aquatic organisms.

Microinjection of Eggs/Embryos

Microinjection is the most popular and well-established method of gene transfer in fish because of its reliability. One reason is that fish eggs are about 1000-30000 times larger than mammalian eggs, which makes injection into the embryo or yolk fairly easy, an experienced person can inject more than 1000 embryos per hour. The gene constructs that were used in this technique include human or rat growth hormone (GH) gene, rainbow trout or salmon CHcDNA, chicken 8-crystalline protein gene, winter flounder antifreeze protein gene, *Escherichia coli* (3-galactosidase gene and *E. coli* hygromycine resistance gene. In general, gene transfer in fish by microinjection is straight forward but technically tedious. 1) number of oocytes or fertilized eggs in embryonic stage are collected from natural spawning or from *in vitro* fertilization; 2) the individual oocyte or egg is held in position by an especially designed apparatus. The injection apparatus consists of a dissecting stereomicroscope and two micromanipulators, one with a microglass needle for injection and the other with a micropipette for holding fish oocyte or embryo in place. Routinely, about 10-10 molecules of a linearised transgene in about 20 ml are injected into the oocyte or embryo, which is then left to grow. There are however, variations in the procedure depending on the fish species involved and the laboratory from which the protocol originated.

Electroporation

Electroporation is a successful method for transferring foreign DNA into bacteria, yeast, plant and animal cells in culture and is considered as an efficient and versatile massive gene transfer technology. This method has become popular for transferring foreign genes into fish species, such as, salmon, trout and carp, which have limited seasonal availability of very large numbers of eggs. Electroporation involves placing the eggs in a buffer solution containing DNA and applying short electrical pulses to create transient openings of the cell membrane, allowing the transfer of gene material from the solution into the cell. However, the exact behaviour of the cell membrane under the influence of electroporation is not known. The efficiency of the electroporation is affected by a variety of factors including voltage, number of pulses and frequency of pulses. An alternative electrical approach has also been reported. It involves use of thin film electrodes on a glass plate to apply localized electric field to the animal poles of fertilized eggs.

The first successful gene transfer by electroporation was demonstrated on medaka fertilized eggs by Inoue *et. al.* (1990). Powers *et al.* (1992) then demonstrated that electroporation can be more efficient than microinjection and is an excellent method for the transfer of DNA into a large number of fish embryos in a short time period. Higher rate of integration, sometimes as much as 30-100 per cent were obtained using electroporation rather than microinjection of DNA.

Sperm-Mediated Gene Transfer (SMGT)

SMGT has been used to produce transgenic fish (Khoo *et al.*, 1992). This method involves culturing of viable sperm with exogenous DNA. DNA binds to the post-acrosomal region of sperm and is rapidly internalized. Delivery of bound exogenous DNA into the oocyte and ultimately the oocyte genome occurs as a result of fertilization.

Gene Gun Technique

It involves the firing of microscopic particles coated with DNA constructs directly into the living cells, through the use of devices known as gene guns (Lutz, 2001). The particles themselves, usually made of gold or tungsten, carry the DNA into the cell interior where it can subsequently be incorporated into the host genome. This method has also been adapted for gene transfer in fish eggs. In this method, fertilized eggs of loach, rainbow trout and zebra fish were bombarded with high velocity tungsten microprojectiles covered with DNA. Transgenic individuals were obtained (Zelenin *et al.*, 1991). This method, however, though simple, requires expensive and sophisticated apparatus.

Characterization of Transgenic Fish

Once a transgene construct is introduced into the cytoplasm or the nucleus of a fish egg, the foreign DNA can undergo either of three fates, namely, (i) it can be eliminated through the nuclease digestion, (ii) it might get incorporated into the genomic DNA, or (iii) it may recircularise, concatamerise or remain as a free copy DNA in the cytoplasm.

If it gets integrated into the host genome, it may occur before the first cleavage division or after one or several rounds of cell division. Even if it gets integrated, it may or may not express itself and may or may not perpetuate through germline transmission (Padhi and Mandal, 2000). The factors determining sites of integration are still poorly understood. It is particularly important to gain greater accuracy of controlled site of integration because of the unpredictable effects of uncontrolled integration on resident genes.

Expression of Transgenes

The uptake and integration of a transgene does not guarantee that the gene will express itself in the new genetic environment (Ayyappan and Gopalakrishnan, 2006). Tests are to be carried out to determine whether there is expression and if there is expression, at what level this takes place.

Depending on the levels of transgene products in the transgenic individuals, the following methods are commonly used to detect transgene expression; (i) Northern or dot blot hybridization; (ii) RNase protection assay; (iii) Reverse transcription-Polymerase chain reaction (RT-PCR); (iv) Immunoblotting assays and (v) Other biochemical assays for determining the presence of the transgene protein product.

Among these assays, RT-PCR is the most sensitive method and only requires a small sample. Recently, Tanaka *et al.* (2001) used the green fluorescent protein

(GFP) to monitor stable gene expression in medaka. By using GFP as a reporter gene, patterns of expression in embryos and adults can beeasily monitored *in vivo* under a fluorescence microscope.

Inheritance of Transgene

Inheritance study is useful to know whether the transgene is transmitted to the next generation. To determine this, the transgenic individuals are mated to nontransgenic ones and the progeny are assayed for the presence of transgenes. Detailed analysis of the F1 and F2 generations in a number of transgenic individuals would establish stable incorporation of the gene constructs into the host genome. Direct evidence for transgene integration may require localization by *in situ* hybridization.

Application of Transgenic Fish

Three aspects of fish growth could be enhanced for economic purposes; (i) initial growth rate such that they reach maturation earlier, (ii) enhanced growth rate as adults to provide larger fish for market and (iii) fish with improved feed efficiency. GH sequences driven by non-piscine promoters elicited growth enhancement in transgenic carp, catfish, zebrafish and tilapia. GH genes are reported to grow 10-80 per cent faster than non-transgenic fish in aquaculture conditions if the proper promoters are utilized.

The anti-freeze protein gene from winter flounder (*Pseudopleuronectes americanus*) has been isolated and cloned. This species avoids freezing of its blood even at sub-zero temperatures by producing a set of anti-freeze proteins. The primary purpose to produce salmon that could be farmed under Arctic conditions, but expression levels obtained have been inadequate for increasing the cold tolerance of salmon.

By application of transgenic techniques in fish it is possible to produce sterile fish by antagonizing the action of hormones such as gonadotropin releasing GnRH of gonadotropin GtH.

Mammalian genes (MHC, T-cell receptor, immunoglobulin, lymphokines) implicated in regulating disease are currently seen as possible candidates for gene transfer experiments for enhancement of disease resistance for viral and bacterial diseases.

Objections for Production of Transgenic Fish

The techniques used to produce the transgenic fish are unethical and therefore undesirable. The transgenes used could offer undesirable as well as desirable new traits on the genetically modified fish. Transgene incorporation could lead to genetic problems through position effects or other genetic interactions. The novel proteins in the transgenic fish could evoke immunological and other medical tolerance problems in consumers. The transgenic fish might interbreed in the wild with native fish leading to negative impact on environment. The fish, although not interbreeding in the wild, could be viewed as an undesirable new alien species analogous to an introduced species.

Chapter 16

Genetics of Common Carp and other Food Fishes

Among the edible fishes, the carp has been the object of most of the studies undertaken so far on genetics of morphological and physiological characters. There are some data available on the genetics of Crucian carp (*Carassius auratus*), but these are concerned chiefly with the inheritance of differences between the aquarium races of the gold fish- a domesticated variety of the crucian carp. The information available on other edible fishes (mainly on the inheritance of single trait) is fragmentary.

Consideration of all categories of hereditary differences found in fish enables one to distinguish four basic groups of traits, namely;

1. Large qualitative morphological-anatomical differences, which are distinctly segregating and rather independent of environment. Hereditary differences is colour, and scale patterns of the common and crucian carp belong, in particular, to this category;

2. Quantitative differences in morphological and physiological traits, which are inherited polygenically and affected by environment to a great extent. In this case the expression of the character is determined by the combination of many genes adding to each other (additive action) and to environmental conditions, variation of body weight, vertebrae number, number of fin rays, variation of many physiological traits etc., come under this category of hereditary differences in fish.

3. Biochemical differences which are expressed in the variation of blood groups, presence of different forms of haemoglobin and transferring in blood, the presence of some allied forms of enzymes (isozymes) etc. The differences in all these indices are inherited quite distinctly and are

determined by a small number of completely segregating genes. In the majority of such cases codominance is observed- each homozygote is provided with one protein corresponding to the given gene, whereas the heterozygote has both protein characteristic of homozygotes. Usual complete dominance is rarely found.

The new method for the detection of fine distinctions of protein molecules (electrophoresis, serological reactions, transplantation of organs etc) enable research of the genetical variability of a population without crossings and hybrid analysis Direct biochemical examination of the great number of Individuals in any population provides the knowledge of the genetic structure of the population. When applying this method one studies not the specimen's phenotypes but the initial or primary products of gene activity.

4. Phenodeviants, which are deviations (in structure) and malformations found in many domesticated and natural fish populations. Both the number of phenodeviants and degree of deviation grow under inbreeding and unfavorable environmental conditions. As a rule the inheritance of phenodeviants is complicated, and the character often expressed incompletely, sometimes apparently returning to the normal structure. Distinct Mendelian relations are seldom observed in these cases.

The types of hereditary differences are found in all fish irrespective of their systematic position. However, it will be more convenient to consider the genetics of separate species rather than types of hereditary peculiarities.

Genetics of Pond Fishes – Common Carp and Crucian Carp

The works of Kirpichnikov and Balkashina (1935, 1936), Golovinskaya (1940), Kirpichnikov (1937) and Probst (1949) demonstrate two pairs of autosomal unlinked genes which determine the type of scaliness in carps. The following genotypes carp are possible.

SSnn, Ssnn – scaled; ssnn – scattered; SSNn, SsNn – linear; leather. Carp with genotypes SSNN, SsNN and ssNN are not viable. N gene is lethal in the homozygous state and embryos die in the hatching stage. This gene has a reducing effect on many organs of heterozygous carp.

The data on the heredity of basic types of scaliness have been corroborated by recent works. It is obvious that S and N genes arose in European carp (*C. carpio* L) as a result of two independent mutations. Later their combination produced the leather carp which is without or almost without scale. There are linear and leather forms among the Japanese carp, *C. carpio haematopterus*. According to K.A. Golovinshaya, the principles for heredity of scale patterns for these carps are identical to that of the European sub-species (*C. carpio carpio*). Trandinh Trong (1967) and Kirpichnikov (1967a) have described the mirrior and linear carps of Vietnam (Sub-species of *C. carpio fossicola*). Thus, analogous scale genes have arisen among carp belonging to three sub-species which differ quite distinctly. This may serve

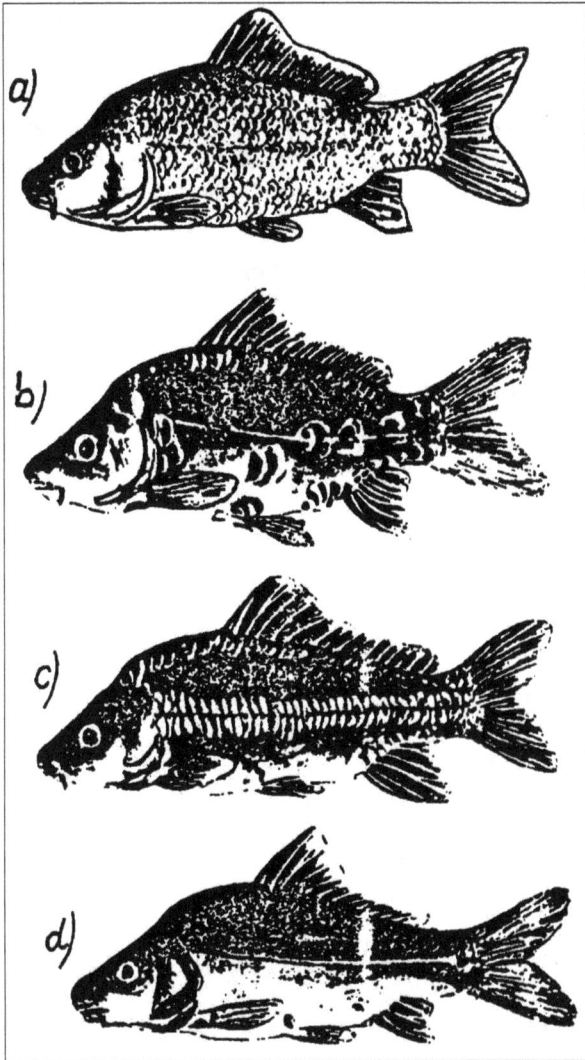

Figure 16.1: Types of Scale Cover in Common Carp.
(a) Scaled; (b) Scattered; (c) Linear; (d) Leather.

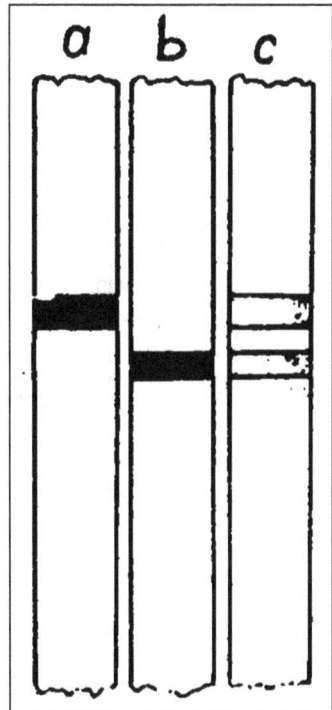

Figure 16.2: Electrophoregram of
Haemoglobin in Fish (Scheme).
(a and b) Homozygous; (c)
Heterozygous genotypes.

as a striking example of how Vavilov's law on homologous hereditary variation is expressed in fish (Vavilov, 1935).

The table below gives the results of all possible crossings of carp having different types of scale patterns. Segregation corresponds to the expected pattern; but the number of linear and leather carps is low due to their lower viability. At lower mortality rates, the correspondence of the actual and theoretical frequencies may be closer.

Table 16.1: Scale Cover Inheritance in Common Carp

Sl.No.	Parents (Irrespective of sexes)	Number of Offsprings (in per cent)			
		SC	St	Lin	Leath
1.	Sc x Sc	100	–	–	–
		75	25	–	–
2.	Sc x St	100	–	–	–
		50	50	–	–
3.	Sc x Lin	50	–	50	–
		37.5	12.5	37.5	12.5
4.	Sc x Leath	50	–	50	–
		25	25	25	25
5.	St x St	–	100	–	–
6.	St x Lin	50	–	50	–
		25	25	25	25
7.	St x Leath	–	50	–	50
		33.3	–	66.7	–
8.	Lin x Lin	25	8.3	50	16.7
9.	Lin x Leath	33.3	–	66.7	–
		16.7	16.7	33.3	33.3
10.	Leath x Leath	–	33.3	–	66.7

Sc: Scaled; St: Scattered; Lin: Linear; Leath: Leather.

Golovinskaya (1946) crossed a female linear carp with two leather males and obtained the following:

	SC (Sn)	St (sn)	Lin (SN)	Leath (sN)
Expected	725	725	1450	1450
Actual	758	758	1406	1426

Two heterozygous linear spawners, when crossed, yielded the following segregations (Wohlfarth, Lahman and Moav, 1963).

	SC (Sn)	St (sn)	Lin (SN)	Leath (sN)
Expected	301	100	602	201
Actual	343	109	568	184

In both cases, the number of linear and leather fingerlings was less, whereas the scaled and scattered ones were more numerous than expected. The lower viability of carps with N gene is due to the pleiotropic effect of this gene. Alleles of another series, S and s have the pleiotropic effect as well; many organs of carps are affected by the scale genes and many morphological and physiological characters are changed. Most striking is the difference between carp with or without N gene (the linear and leather, the other scaly and scattered). Among the characters which are affected by N gene, the one to be mentioned particularly is the number of pharyngeal teeth, as can be seen in linear and leather carps.

Table 16.2: Pleiotropic Effect of Scale Genes in Carp

Character	Type of Carp				Source
	Sc S,n	St s,n	Lin S,N	Leath s,N	
Weight of fingerlings optimum conditions	100	93-96	85-88	79-80	2,3,4
Weight of fingerlings depressed conditions[2]	100	83-94	42-70	37-72	2,3
Weight of 2-year-old fish[2]	100	94-96	86-91	83.84	3,4
Average no.of soft rays in dorsal fin (D)	18.8 (17-22)	18.7 (17-22)	16.4 (12-19)	15.4 (15-18)	1,2
Average no. of soft rays in in anal fin (A)	4.96	5.00	3.82	3.56	2
Average no. of rays in ventral fin (V)	8.91	8.68	8.76	8.47	2
Average no. of soft rays in pectoral fin (P)	14.7	14.3	14.3	13.1	4
Average no. of gill rakers (variation of means)	24.6-25.1	24.3-24.8	19.4-21.6	18.5-20.5	1,2,4
Number of gill fringes	88.6	83.5	82.3	83.2	4
l/H ratio (index of elongation)	2.77-2.33	2.74-2.26	2.86-2.35	2.82-2.35	4
Average no. of pharyngeal teeth	9.22	9.58	7.63	7.44	2
Regenerative capacity of fin[2]	100	76	39	19	2
Length ratio of back to front chamber of swimming bladder	>1	<1	?	?	1,2

Contd...

Table 16.2–*Contd...*

Character	Type of Carp				Source[1]
	Sc S,n	St s,n	Lin S,N	Leath s,N	
Heat resistance (critical temperature, C degree)	37.6	37.5	36.8	36.6	5
Resistance to oxygen deficiency survival in minutes	210	210	132	132	5
Erthrocytes number in millions/cm cube	1.93	1.99	1.76	1.69	5
Haemoglobin, g/per cent	9.02	8.87	8.18	8.28	5
Viability of fingerlings optimum condition	100	91-98	87-93	80-92	2,3
Viability of fingerlings depressed condition	100	93-95	36-37	28-60	2,3

*: 1. Golovinskaya, 1940; 2. Kiprichnikov, 1945, 1948; 3. Probst, 1953; 4. Steffens, 1966; 5. Chan mai-Tohen, 1968.

2: Expressed as percentages of the values of incidence of the same character in scaly carp.

Sc: Scaly; St: Scattered; Lin: Linear; Leath: Leather carp.

Instead of usual three rows of pharyngeal teeth (1.1.3-3.1.1) often two or only one row of them (1.3-3.1; 1.3-3 and 3-3) respective was found.

According to Probst (1953) the reduction of many organs observed in linear and leather carp is accompanied by a defect in the mesenchyme development. There is doubt that N gene represents a major mutation (perhaps of the deletion type) completely destroying the synthesis of one or more proteins of vital importance. This may explain the decrease in vitality in heterozygotes with N gene and mortality of all homozygotes. The difference in the structure of the swimbladder is to be considered a result of one of the pleiotropic effects of a gene. When a gene is present, the back chamber is shorter than the front one (scattered carps); with normal allele S it is, on the contrary, longer (scaly carp).

Carp with different scale genes may be distinguished by means of serological methods as well. Aitouchov *et al.* (1966) and Pochil (1967) have been able to differentiate between all the four genes by means of erythrocyte antigens.

The colour of carp is a character inherited distinctly and, in the majority of cases, simply. Most numerous and varied are the types of colours found in carps of Japan, Vietnam and Indonesia.

The heredity of blue colour of Polish carp has been analysed. A ratio near 3:1 has been obtained for the second generation, the number of blue fingerlings varying from 20.3 to 23.8 percent (Wlodek, 1963). Blue carp were characterized by somewhat lower viability. They at first grow better than common carp and then lag behind them. For the German blue carp the ratio of F2 and F1 (back cross) proved to be near to that expected (3:1 and 1:1 respectively) and no difference in viability were found.

Blue Israeli carp (b gene), similar to that Polish ones, were inferior to non-coloured carp both in viability and growth rate. In all the three cases the mutant form was recessive. Two more recessive mutations for colour-gold (g gene) and grey (gr gene) were found in Israeli carp. In one of them, all were homozygous by g gene and in another by b and gr genes. Although both lines fell behind the non-coloured ones in viability and growth rate, the crossings between them produced non-coloured (yellowish) carp with high fishery management indices.

$$g/g \quad Gr/Gr \quad B/B \quad X \quad G/G \quad gr/gr \quad b/b$$

gold   grey, blue

$$G/g \quad Gr/gr \quad B/b$$

Non-coloured (heterozygotes)

All three genes proved to be autosomal and not linked with each other.

There are many coloured variations of carp in Japan, including yellow, blue, red, red and white, spotted etc.The heredity of these types of colour not yet been analyzed, but it is known that they are determined by both dominant and recessive mutations. Studies of the red, brown, violet, black coloured variations of Vietnam and Indonesian carp, and carp of China, Korea and other countries of east and south Asia are needed. Similar mutations of colouring arise among carp inhabiting various countries and belonging to various sub-species. Thus homologous variability is being presented by this character also.

Quantitative differences, inherited polygenically, compose the second rather numerous category of hereditary differences in carp. The best study is that concerned with the heredity of the number of rays in the dorsal fin and vertebrae number. There are some data on the heredity of other characters including growth rate and body shape. While the crossing one female having 20 soft rays in the dorsal fin with different males, correlation between the number of rays in the offspring and in the males proved to be low but quite reliable ($r = 0.326\pm0.021$). The correlation was more significant ($+0.76\pm0.07$) when the cultured carp, Amur wild carp and the hybrid between the two (various generations) were compared. For one generation (for example the third hybrid generation) the correlation coefficient is slightly lower ($r = +0.60\pm0.14$).

The inheritance of vertebrae number has a polygenic character as well. Six crossings were analyzed, which was subdivided into two groups. The number of vertebrae in carp was determined by large number of genes and was dependent to a great extent on environment. According to Nenashev (1961) the heritability coefficient of these two characters is equal for carp (0.4-0.6), that is, the values of the environmental and genotypical effects are approximately equal.

The body height of carp is inherited polygenically as well. Israili breeders selected carp with different height indices (H/I) Offspring of these carp had the following means (Moav and Wohlfarth, 1967).

Table 16.4

Body Weight Index (H/l percent)	Height in Percent of Length		
Parents	Progeny		
	I	II	III
35.8	32.8	35.3	34.4
38.8	33.7	36.7	35.3
43.7	35.6	39.2	37.8

Realised heritability proved to be equal to 0.42. Thus the genetic differences between the various carps were rather considerable in this case as well.

Growth characters (weight and length increments) are most important characters in this group of traits. The heritability of body weight of carp fingerlings are not high (0.1-0.2) and the influence of environment on this character is rather great. Heritability indices for weight in carp of older age are somewhat higher, but this exact value are not stated. Selection for weight is effective in the negative direction but at the same time, rather often there is no manifestation in the positive direction. The differences in growth rate are inherited non-additively. It is quite probable that there exist, in carp populations heterozygous systems balanced for this character, and heterozygotes grow as a rule, better than homozygotes. Inbreedings, responsible for the increase of homozygosity in carp, brings the drop in its growth rate.

Numerous physiological indices are to be considered among the characters of carp inherited polygenically; but no data are available so far on the mechanism of their heredity. There is no information on the heredity of fine biochemical differences as well. The existence of polymorphism of transferring has been found. When crossing common carp with crucian carp the differences in the characters of esterase are inherited codominantly.

Phenodeviants have been discovered in many highly inbred populations and stocks of carp. While selecting the Ropsha carp it was observed, as a result of some crossings, the appearance of mirror carp characteristic of a peculiar coating on the body and an extremely low growth rate.

Malformation of vertebral column, various degrees of reduction of gill cover, deformation of fins etc are recorded for many families of carp. The deviations of one type vary from 0.1-0.2 percent to 20-30 percent or more, the frequency depending on the degree of inbreeding and the environmental conditions under which the fry are reared. In all these cases there exists a certain hereditary predisposition to malformations, to be more exact, they are determined both by environment and genotype.

No gene linkage has been found in carp so far. The mechanism of sex determination has not yet been determined either. This is due mainly to the large number of chromosomes in carp (2n = 104).Diploid number of chromosomes of many cyprinids is equal to 50-52, which allow to consider the common carp as a natural (very ancient) tetraploid.

Most essential studies on genetics of gold fish have been carried out by Japanese and Chinese scientists. Chen (1928, 1934) has shown that the blue and brown colours, characteristic of some strains as well as transparency of cutaneous covering, are inherited by means of a small number of genes. Absolute and mosaic transparency are determined, according to Matsui (1934b) by the combination of two pair of genes. Semi-dominant T gene (absolute transparency) is epistatic in homozygous state to recessive n gene (from another pair) causing frame location of coloured and non-coloured sections. Eyes of gold fish may be telescopic due to the presence of one recessive gene, while elongated, bifurcate fins arise in gold fish as a result of interaction of not less than two or three pairs of genes.

Genetics of Edible Wild Fish

Striking quantitative differences have been found in many species of wild fishes, but only in a few cases they were analyzed from a genetic point of view.

The heredity of vertebrae number in *Zoarces viviparous* depends on the combination of many genes. Each local population of *Z. viviparous* is characterized by a certain average number of vertebrae fixed by heredity. The hereditary nature of differences in the number of gill rakers between the white fish populations of the genus *Coregonus* has been observed. In all freshwater and sea fishes large groups of quantitative characters inherited polyfactorially.

More data are available on biochemical hereditary polymorphism of natural fish populations. The presence of two or more blood groups in one population has been demonstrated for the herring and pilchard (Cushing, 1964), Norway haddock (Sindermann, 1962), anchovy and horse mackerel (Limansky, 1964). Large variation of blood groups is characteristic of different salmonid species. Californian gold trout (*Salmo aguabonita*) has shown individual and population differences in blood groups. At the same time, no considerable serological alterations were recorded for the species during the 90 years which passed after acclimatization of that trout in a water body new to it. An inbred trout population in Summit lake, Nevada, is immunologically homogenous, while pond rainbow trout have several serologically distinguishable groups.

Each population of *Katsuwonus pelamis* L., *Thunnus germo* Gm and other Pacific tuna proved to be polymorphous with regard to one, two or thee erythrocyte antigen systems. Most typical of tuna (as well as other fishes) are 3-allele systems of ABC type similar to the same system of blood groups in man; the latter are characteristic of the presence of four serologically distinguished phenotypes (A,B,AB and O). Various populations differ distinctly in the proportion of phenoyupes.

A most important result of this field of research is the evidence of the presence of wide intrapopulation polymorphism for blood groups in fish.

The application of the electrophoretic method permits the study of genetical variation of fish for haemoglobin, transferrins, muscle and serum proteins.

Polymorphism for haemoglobin genes has been discovered in other edible fish as well. Crossing of allied fish species, the hybrids show the hybrid haemoglobin in addition to parental ones. The haemoglobin molecule of both fish and higher vertebrates is a polymer. Hybrid haemoglobin was discovered by Manwell *et al.* (1963) in offspring from crossing between the different species of sunfish, and by Sick *et al.* (1963) while crossing two species of plaice.

General polymorphism for transferrins has been recorded for cod and four species of tuna. Tuna are polymorphic for serum esterase. Three species of tuna have been studied and were found to provided with two, three and four alleles of esterase respectively. Herring proved to be variable for lactate dehydrogenase and aspartate aminotransferase; roach for esterase, wall eye (*Stizostedion vitreum*) for myogens and trout for lactate dehydrogenase.

Apparently all fishes are characterized by the large genetic variation for protein structure, while many allele systems are in equilibrium. The electrophoretic method is especially convenient for the study of genetic structure of a population due to the simplicity of detection of heterozygotes (the presence of two bands on the paper). The heredity of biochemical differences in fish is primarily codominant. Codominance is a result of the presence in heterozygotes of two such products instead of one.

Additional information on the biochemical polymorphism of fish population may be provided with the experiments on tissue transplantation. This method proved to be advantageous when analyzing the degree of homogeneity of small populations (for example local stocks of reds and other anadromous Salmonidae). A comparison was carried out between heat resistance of muscles and muscular proteins of some races of salmonid fish (*Coregonus autmnalia* and *Thymallus arcticus*) of Baikal Lake and that of Norway haddock from different parts of the northwestern Atlantic. Intraspecific differences have been discovered for this trait which also may be considered as biochemical polymorphism.

After the discovery of the biochemical polymorphism of genetic origin in fish the question arose about the factors responsible for it. The following explanations were put forward;

1. Polymorphism is connected with the selective preference of heterozygotes for the single gene (Ford, 1966);

2. Polymorphism is a result of an alternating (in time and space) adaptation of two closely relate but biochemically distinguishable, homozygotes (AA and A1A1) or homozygote and heterozygote (AA and AA1). In this case, the first genotype is better adapted to one environment, and the second one to another environment (Timofeeff-Ressovsky and Svirezhev, 1967).;

3. Polymorphism is transient-in the past, individuals of one genotype were better adapted, but at present, another one is better (one allele has been substituted for another).

The first two mechanisms are found in fish. Examples of transient polymorphism are not known for fish at present. The application of new methods for genetic analysis of fish populations inhabiting natural water bodies resulted in quick development of research on fish genetics and selection. Laws for the inheritance of many variable characters, especially biochemical peculiarities, have been discovered. The most recent genetic works provided with valuable information on the intraspecific differentiation in fish, nature of differences between populations and variation within populations. Many species seemed to consist of sub-species, or smaller systematic groups which proved to be sufficiently isolated and well distinguishable genetically.The introgression between those groups was not large in the majority of cases.

The variability of chromosomes based on so-called Robertson translocations is frequently found in fish. (Simon, 1963; Roberts, 1964). However, the variation of chromosome number occurring in case of such translocations is not accompanied by the change of the number of arms, and it is apparent that the amount of DNA in a chromosome set does not vary either. Sometimes this type of difference in chromosome number is observed within one population, in others two different populations may be distinguished by this character. Karyological variability of this type is to be considered as one more highly essential form of hereditary differences in fish.

Chapter 17

Genetic Method of Fish Selection

Significance of Crossing in Fish Selection

Artificial selection and crossing are the principal means of fish improvement. Among different methods of the utilization of crossing in fish selection, two principal ways of crossing can be mentioned.

(a) Commercial crossing directed toward breeding of the first generation hybrids for commercial purposes. The importance of commercial crossing in the fish culture industry is rapidly growing. In crossing of this kind only the first generation hybrid is used, characterized by a heterosis of productive qualities or incorporating the advantageous characteristics of both the parental forms. First generation hybrids, as a rule, are not preserved for further reproduction.

(b) Synthetic crossing (of various degree of separation, including intergeneric) whose purpose is the development, in the course of long selection, of a new breed. It aught to combine the best qualities of the parents (of 2, 3 or sometimes 4 breeds, 2 or more species or even 2 genera of fish). A particular pattern of crossing in new-breed production should ensure the preservation and perfection of the production qualities of the breed, the preservation of genetic variability, and the prevention of inbred depression. Lately, use has been made of such crossing patterns that allow heterosis to be utilized to its utmost in every generation.

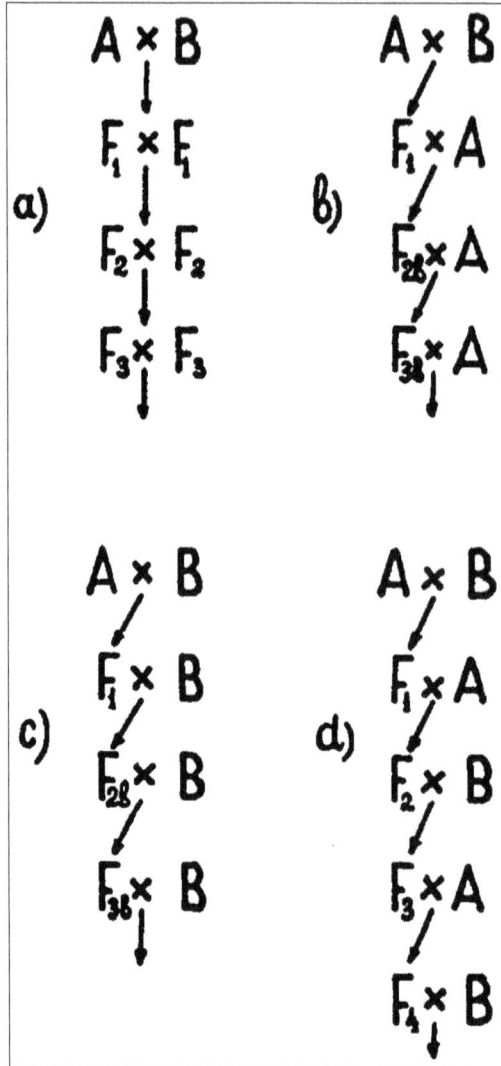

Figure 17.1: Types of Crossings in Synthetic Selection.
a: Reproductive; b: Introductive; c: Absorptive; d: Alternative.

Inbreeding and Out Breeding in the Fish Culture Industry

1. Inbred Depression

Inbreeding is used in large scale in a wide variety of cattle rearing and plant cultivating industries. It can be used in fish selection as well.

It is known that the inbreeding measure is the coefficient of breeding incorporating the degree of the animal's homozygosity. It shows what part of the genes in a group of individuals(expressed in fraction of one or in percentage of the total genes in a haploid combination) are in the homozygous state (Falconer,

1960). Close inbreeding, especially, sib-mating (brothers with sisters and parents with children), causes homozygosity to rapidly increase. Accordingly, the coefficient of inbreeding also increases up to 0.90-0.95 and more (sometimes as high as 1). In most animals with different sexes,inbreeding results in a drop in viability, rate of growth and quite often in fertility. This deterioration of inbred specimens is termed as inbred depression. Inbred depression was found in studies of many fish. A few are enumerated below.

Carp (*Cyprinus carpio*) -> A marked drop in the viability and rate of growth of carp during inbreeding has been recorded by many selectionists (Kuzema, 1950, 1953; Shaskolsky, 1954; Lieder, 1956; Kirpichnikov, 1960, 1961, 1966; Schaperclaus, 1961; Golovinskaya, 1962 and many others).

The problem is dealt with in detail by Moav and Wohlfarth (1963, 1967, 1968). According to their data, full-sibling mating (brothers and sisters) decreases the growth rate of commercial carp by 10-20 percent. At the same time there is a drop in viability, and the number of anomalous offspring abruptly increases. According to Kirpichnikov, 1969, even moderate inbreeding, used in the selection of the Ropsha carp, resulted in their deterioration. In certain families, many deviations from normal were observed, the so-called phenodeviants.

Inbred depression in carp may have a very pronounced manifestation. Estonian carp, for example, subjected to inbreeding for a number of years, have a growth rate of only one-half that of the Ropsha carp.

In accordance with the degree of inbred depression, out breeding is accompanied by heterosis in growth rate and viability. Especially considerable is heterosis, when the fish from different highly inbred groups are crossed. The higher the coefficient of inbreeding in these groups before crossing, the greater the reduction of the coefficient in the outbred offspring. The considerable manifestation of inbred depression in carp can be attributed to the peculiarities of its biology and the structure of its chromosome system, high fecundity (over1000000 eggs), large natural populations, as well as the high natural heterozygosity in this species. The reduction of heterozygosity in these conditions, owing to inbreeding, would always be detrimental.

Common Trout and Brook Trout

As regards the salmon family, the data on the effect of inbreeding are scarce and discrepant. In brook trout (*Salvelinus fontinalis*) the occurrence of inbred depression was established by Cooper (1961). In inbred lines, the growth rate of fish was retarded. However, in the experiments conducted in Ropsha in crossing trout of different origin, heterosis has not been detected. This evidence seems to indicate that there is no inbred depression in these trout. In evaluating the possibility of inbred depression in man-bred salmon, it should be borne in mind that their fecundity is relatively low (up to 2-3 thousand, rarely up to 5-10 thousand eggs from one female). The fish breeder has to keep a great number of spawners, frequently in the thousand range. The number of fish being so great, the hazard of the coefficient of inbreeding increase is insignificant, and the population as a whole remains quite heterogeneous. This fact can easily account for the absence of marked heterosis in crossing different groups of trout. In close inbreeding effected by the selectionist, inbred depression may easily result; this depression however, is eliminated in crossing.

Data pertaining to other fish are very scarce. Gibson (1954) points out that in the guppy (*Lebistes reticulates*) there is a drop in viability resulting from inbreeding. Krivoshchekov (1963) believes that the reduction in the size of crucian carp in small lakes can be attributed to inbred depression, resulting from inbreeding with small populations of crucian carp. There are scattered data pointing to the detrimental effect of inbreeding in other kinds of fish. The breeding of many kinds of aquarium fish is accompanied by very close inbreeding. The inbred depression in this case can be overcome by a careful selection of best specimens. This peculiarity of aquarium fish is explained by their low fecundity and the occurrence in nature of a great multitude of small isolated populations. In small populations inbreeding is inevitable. The species get adapted to the low heterozygosity, and for this reason inbred depression manifests itself to a very insignificant extent.

Fish of carp and salmon families, bullheads, sunfishes, tilapia and other commercial fish are widely cultivated at presnt. Considerable inbred depression is to be expected in all these fishes in pond breeding. Crossing patterns used in selection should be designed to overcome the depression and to make maximum use of heterosis.

2. Crossings

In fish selection, as well as selection of other animals, distant outbreeding is indispensable. The purpose of such crossings may be described thus;

   (a) An overall increase of genetic variability results in an increase of artificial selection response.

   (b) The achievement of a combination of characters of two or three breeds or two (rarely three) species.

   (c) The improvement of the productive quality of the local breed by making use of the few valuable traits of another breed (improver).

   (d) The increase in the viability of the breed by introducing genes responsible for resistance to environmental factors and diseases.

In accordance with the above objectives, following the original crossing, the reproduction of the hybrid population is carried out by means of reproductive, introductory, absorptive or alternating crossing.

Reproductive Crossing

This method of crossing is carried out when many valuable properties from both parents are to be combined in the hybrid. It can be easily effected with complete fertility, of the hybrids and requires only a meticulous selection in the subsequent generations. In fish breeding, a few instances of successful reproductive crossing is known. With its help Ropsha North carp has been obtained, the Ukrainian breeds of carp have been reared and the work is being carried out to rear the Mid-Russian carp. There are reasons to believe that the pre-World War II Hungarian carp was the result of three-way crossing of the local less productive breed with the german

Aischgrund carp and the carp brought from Japan. For further reproduction of the hybrids reproductive crossing must be used. It was also applied in the rearing of rainbow trout (after the crossing of two or three American trouts) as well as in similar work with some other fishes.

Introductory Crossing

This system can be used to advantage when one or a few characteristics from a breed must be incorporated in the hybrid. Each generation of the hybrid must be crossed with the local breed of fish. The hazard of loosing the useful characteristics of the improved breed in local crossings of this kind is very high. If the characteristics are determined by one or several clearly segregating genes, the problem is solved with little difficulty. In the case of polygenic inheritance of the properties selected, the selection difficulties are sometimes impossible to overcome.

There has been no indication in the literature that introductory crossing has been put into practice. Its application has been envisaged in the selection of Israeli carp in two cases; in the crossing of the local highly productive carp with those brought from Netherlands and in the making of breed strains marked by the colouring genes. The first introductory crossings have already been carried out with the object of introducing colouring markers into the best strains of the Israeli carp (Moav and Wohlfarth, 1967). The markers in this case are three recessive genes, determining gold, blue, and grey colouring of the fish.

Introductory crossings has a lot of potentialities when the selectionists attempt the task of increasing a breed's resistance to a certain disease. Resistance to disease often depends on the presence of one or few genes, and they can be preserved by means of meticulous selection in each generation. Selectionists have succeeded in rearing breeds of fish that will have high resistance to some of the most dangerous diseases.

Absorptive Crossing

This resembles introductory crossing. The only difference is that the purpose of this type of crossing is a nearly complete substitution of the local breed of genotype by the genotype of the improved breed. Only some peculiarities typical of the local breed must be preserved, pertaining mainly to its viability.

In both introductory and absorptive crossing, selection is seriously handicapped if the high viability of the local breed is determined by a large number of genes.

As new high-quality fish breeds are reared, the significance of absorptive crossing in commercial fish breeding will increase.

Alternate Crossing

This method requires the most complicated system of rearing, and so far it has not been used in commercial fish rearing. It is especially advantageous when a combination of many characteristics from two breeds with polygenic inheritance is required. Alternate crossing provides for the preservation of high genetic variability through a number of generations. Owing to this variability, selection efficiency is kept at a high level and selection plateaus do not result.

After several (3, 4 or more) alternating crossings, they must be replaced by the conventional reproductive crossings. Otherwise stabilization of breed characteristics will be difficult to achieve.

With alternating crossing, the coefficient of inbreeding practically does not increase. This enables specimens from each hybrid generation to be used directly for commercial breeding.

Undoubtedly alternate crossing and in more complicated forms will be used in the future fish selection. However, it will require experienced selectionists and well-equipped selection stations. If the selectionist faces the problem of obtaining new hybrid breeds by means of crossings (interspecific or intergeneric), the principal difficulty in hybrid selection of their complete or partial sterility. The restoration of normal fertility takes a lot of hard and time-consuming work. The results of Kuzema and Tomilenko's (1965) experiments with common and crucian carp hybrids indicate that back crossing of the hybrids with one of the original species may help to restore the fertility in the hybrid form.

Different Systems of Breeding

To utilize completely the advantages associated with heterogeneous crossings, fish breeding should be carried out according to a rigid pattern.

Parallel Breeding of Two or more Groups

While working with slowly maturing fish, it is practicable to have two or three groups concurrently within a breed, without intermingling, allowing inside each a moderate inbreeding; and carrying out selection in each generation. For commercial purposes, fish from different groups are crossed. The prevention of close inbreeding enables one to avoid considerable exhaustion of hereditary variability within the breed. This method is used for the selection of Ropsha carp.

Breeding in Groups with Family Selection

In each generation a large number of crossings of fish from different groups is carried out (in Israel up to 20 combinations a year). The parents producing the best offspring are used for subsequent commercial crossings. For further reproduction those offsprings are selected that have maximum overall combining ability. The weak spot of such system of breeding is the gradual narrowing of the genetic variability in carrying out family selection.

Breeding in Groups keeping Reserve Gene Pool

Moav and Wohlfarth (1967) recommended, that in selecting two groups marked by certain genes, a reserve group of fish, numerous enough, shoud be kept for each group. In case of a genetic variability drop, it will allow an additional gene pool to be introduced into the exhausted groups. As in the previous case, selection should aim at an increase in combining capacity.

Alternating Inbreeding and Out Breeding

After two or three generations of close inbreeding, the evaluation of the hybrids from different inbred lines is performed. The best combinations are used for commercial rearing and among the offspring new inbred lines are established. It seems possible to obtain 4-step hybrids. The practicability of such a system in fish breeding must be corroborated by appropriate experiments.

Linear Selection of Superior Type

Linear selection involving inbreeding for superior anscestors was started by A. I. Kuzema in 1950 in his experiments with the Ukrainian carp. The problem in this type of selection lie chiefly in the difficulties of accurate evaluation of spawners and the necessity to choose a few of the best of a great number of fish.

Top Cross

The term top cross is applied to a selection technique whereby one or several inbred lines and a large out bred groups are maintained at the same time. To preserve the genetic variability and to improve the offspring characteristics, the crossing is done between the best inbred specimens (males for instance) and specimens from the out bred populations (females).

Top cross may be applied in commercial fish breeding, especially if inbreeding results in a considerable drop in female fecundity and viability.

Reciprocal Recurrent Selection

Reciprocal recurrent selection being one of the most complicated techniques of animal breeding at present is not likely to find its way to fish breeding. In reciprocal selection, the combining capacity of the parents from each of the two breed groups is evaluated by means of a cross with parents from the other group. The specimens selected as a result, are reproduced without recrossing and their offspring are again tested for combining potential.

The basic peculiarity of all the systems considered is the utmost utilization of heterosis in crossing fish from different groups, lines and breeds. Alongside this, inbreeding is used, varying from quite moderate, in some cases, to very close in others. The future will decide which breeding techniques are most effective in the breeding of pond fish. The correct choice of techniques will be primarily a function of our knowledge in genetics and of the perfection of our knowledge of inbred depression and heterosis in fish crossing.

Special Methods of Selection of Fish

Apart from selection and crossing, special genetic and cytogenetic technique are widely used in present day selection, some of which are applicable to fish.

Direct Application of Genetic Data in Fish Selection

Besides the marking of selected groups of carp by colouring mutations, another example is the back cross study of homo- and heterozygosity in the scaled carp in the genes (mirror and scattered scale).

The back crossing is carried out thus;

1) S(s?) nn x ss nn

   Scaled   Scattered

2) S (s?) nn x Ss nn

   Scaled   Scaled heterozygous

In the former case, the scaled and scattered offspring of the heterozygous parents are produced in a 1:1 ratio. In the latter case, the ratio is 3:1 (three scaled per one scattered). The entire offspring of the homozygous scaled parents in any of the two crossings have a completely scaled integument. Within the period from 1956 till 1964, 469 parents of the Ropsha carp were examined by this technique, 247 turning out to be homozygous. This work enabled the selectionist to eliminate completely, the occurrence of the scattered carp as early as in the 5th selected generation.

The better the genetics of the fish under selection is known, the more data will be at the disposal on the inheritance of certain quantitative and qualitative characters, rendering selection easier and faster.

Use of Radiation and Chemicals for Enhance Mutation

Increase in the number of mutations caused by strong mutagenic actions may considerably increase the population's heterozygosity and selection effectiveness.

The very first experiment with chemical mutagens proves that mutation rates may be multiplied many-fold. Artificial mutagenesis can be used to the best advantage in the selection of pond fish for resistance to certain dangerous infections and parasitic diseases.

Gynogenesis and Androgenesis

These reproduction procedures may be quite useful in fish breeding if it is necessary to rapidly increase the homozygosity of a selection group or line.

Polyploidy in fish is feasible. Sex in fish is determined by sex chromosomes (gonosomes) rather than the balance between sex chromosomes and autosomes. This makes it possible to double chromosome sets in the course of evolution (as it must have occurred in the carp and salmon families). However, the possibility of obtaining practical polyploids in fish is questionable.

Polyploid Fishes: Culture Prospects

A fish normally contains two sets of chromosomes in each of its cells of which one set is contributed by the ovum and the other by the sperm. A polyploidy fish contains more than two sets of chromosomes. Triploid individuals contain three sets, whereas tetraploid individuals contain four sets of chromosomes.

Occurrence of Polyploidy Fishes

Polyploid fishes occur in nature and can also be produced artificially. Artificially produced polyploids may be useful in a) fertility control, b) improving the survival rate in juveniles and c) improving the growth rate in certain species of fishes. Thus, culture of certain species of artificially produced polyploidy fishes can increase the profitability in aquaculture.

Polyploid Fishes in Nature

Polyploidy has been encountered in six different Orders of fishes, but observed more commonly in some paddle fish and sturgeons (Order Acipenseriformes), many salmonids (Order Salmoniformes), all catostomids, some cyprinids and cobitids (Order Cypriniformes).

Naturally occurring polyploid fishes are mostly tetraploids. Salmon, common carp and crucian carp are tetraploid forms. Poecilopsis is a triploid fish found in nature. Five species of fishes belonging to the Order Cypriniformes having high chromosome number have been recorded by Dr. Khuda-Bukhsh and his collaborator (Kalyani University) from the hillstreams of India. The mahseers (*Tor tor*) are most important among them.

Polyploids can be produced artificially by meticulous application of temperature (cold and heat), pressure and chemical shock. Triploids are produced when artificial treatments are made shortly after the fertilization. This prevents the expulsion of the second polar body of the egg. As a result, the offspring produced contains two sets of chromosomes contributed by the female and one set by the male parent. Tetraploids can be produced if the treatments are made shortly before the first cleavage division. The thermal shock treatment is most preferred in polyploidy induction for convenience particularly in the situation when large volume of eggs are to be treated. Triploid offsprings are often produced as a result of hybridization.

By temperature shock triploidy could be produced in stickleback, trout, salmon, channel catfish, flatfish, common carp, grass carp etc. and tetraploidy in tilapia, rainbow trout etc.

Detection of Polyploids

Ployploids can be screened quickly by a) measuring the volume of the nucleus and cytoplasmic ratio in red blood cells and b) by counting the number of nucleoli (sing. Nucleolus) stained by silver nitrate. Flow cytometer and Coultercounter analyzer are the instruments used for quick sorting of polyploids. By determining the chromosome number and correlating it with DNA quantity per cell the presence of polyploidy can be confirmed. Isozyme analysis through gel electrophoresis also helps to detect polyploidy.

Evaluation of Polyploidy Fishes

Growth

The effects of triploidy on growth is a species-specific phenomenon. In Poecilopsis, stickleback and coho salmon the growth rates are similar in triploids and diploids. In juvenile stage coho salmon triploids grow slower than the diploids. Triploid individuals of sexually mature plaice x flounder hybrids are significantly heavier than the diploids. The diploid channel catfishes grow slower than the triploids after attaining maturity.

Fertility

Triploids experience some difficulty in meiosis. So the gonadal development is reduced leading to sterility. The gonadal development is generally lesser in females than in male triploids. Ovary is much reduced in triploid grass carp female. In channel catfish, plaice and plaice x flounder hybrid the diploids have ovaries four times as large as triploids. In triploid and diploid tilapia the testis develops to the same size, but the ovaries of triploids remain markedly underdeveloped. As a result, the female tilapia triploid grows 14 percent faster than the mono-sex male (sex reversed) and 23 percent faster than the triploid males.

Tetraploids are generally expected to be fertile. But this is not so since meiotic process hampers due to multivalent pairing of chromosomes. Often a neuploid gemets, which are not normal, are formed reducing the fertility. Similar to the triploids, the fertility is more reduced in females than in males. Tetraploid male rainbow trout produces milt at the age of two years. Thus, tetraploid males may be crossed with diploid females to yield triploids on a mass scale and the advantageous characters in triploids exploited for commercial gain.

Survival

The survivality is often better in triploids than in diploids. Of all the crosses between rainbow trout, brook trout and brown trout the triploid hybrids survive in a greater percentage than the diploid hybrids. Triploid hybrids of Chinook salmon and pink salmon also exhibit greater survival than the diploid hybrids at the juvenile stage.

Adaptability

Polyploidization may develop greater genomic plasticity which would help the animal to adapt a better way in varying environmental conditions. The common carp, *Cyprinus carpio*, is very labile and adapts nicely in a very wide climatic and limnological conditions for its tetraploid genomic machinery. The catostomids, an exclusively polyploidy group, can thrive well in the diverse conditions.

Cultural Prospects

Triploid grass carp is generally sterile. It is very much useful for controlling vegetation with out creating much concern for environmental damage. The culture of triploid channel catfish is useful in improving the productivity since the sterile ones grow faster than the fertile ones after they attain maturity

Selection Techniques

A few of the most important selection techniques are given below :

Artificial Fertilization

Selection of carp and many other fish with external fertilization is facilitated by using artificial fertilization. In some fish, such as, trout and whitefish eggs and sperms can be obtained without hormone injections. In other fish, such as, carp, crucian carp, grass carp, silver carp and others injection of gonadotrophic hormones is used, mainly the hormone of hypophysis. Ovulation occurs at a certain time after the intramuscular injection of the hormone. In carp this period varies from 10 to 30 hours, depending on maturity and water temperature. After the injection fish are kept in special breeding cages.

Prior to egg taking, large fish should be anesthetized (by quinaldine, MS-222, or other drugs).

The egg and milt are placed into basins or China cups and stirred without adding water (the so-called Russian method of fertilization). If the fish have non-adhesive egg, the fertilization can be performed in a small volume of water. The optimal egg and sperm ratio varies from species to species. In carp it is from 50:1 to 100:1.

After fertilization, adhesive eggs require washing away of the agglutinating agent, and its further secretion must be stopped. Degumming is done by means of two methods;

(a) **Woynarovich's method (Woynarovich and Kausch, 1967):** A solution of a mixture of sodium chloride (4 g per liter of water) and urea (4 g per liter of water) is added to the mixture of egg and milt. One volume of egg requires two volumes of the solution. During first 3-5 minutes the egg is carefully and continuously stirred, fertilization takes place at that time. The subsequent agitation is done intermittently every 2-3 minutes. As the egg swells, additional batches of the solution is added.

In 1-1.5 hours the excess solution is poured out and the egg is placed in another solution (1.5 g of tannin per liter of water) for 10 seconds to strengthen the membranes. The egg should be continuously stirred. Following that, the eggs are washed with fresh water. A little later, it is treated once or twice with a weaker solution of tannin. As a result the eggs are completely degummed and can be placed into Zuger or Weiss apparatus. At the end of this treatment the egg is much larger in size due to considerable swelling.

(b) **The Konradt and Sakharov method (Konradt and Sakharov, 1966):** A solution containing hyaluronidase is added in little portions to the mixture of egg and milt. The hyaluronidase can be easily obtained from the testicles of oxen and pigs. The solution is maintained at a level of 1-1.1 cm above the eggs and the eggs are continuously stirred by a soft feather. In 15-18 minutes, without removing the hyaluronidase, a tannin solution (100 mg per liter) is added to the egg in little portions. The egg is degummed 40-50 miutes after the commencement of the treatment. On making sure that the egg no longer tends to agglutinate, it is transferred to the incubating apparatus.

With the utilization of artificial fertilization, the egg and milt can be divided into many equal parts, the percentage of fertilization can be accurately determined, as well as embryo survival, rate of development etc. All these factors make artificial fertilization indispensable in selection, and especially in evaluating the reproductive capacity of the parents.

Marking of Fish

Quite a number of marking techniques have been tested. The best results obtained were by cutting the fins of fish, branding signs on the skin and scales by a red-hot wire, or subcutaneous injections of India ink and some organic dyes (dichlorotriazine and other compounds). For certain fish, suspended tags can be used. Marking makes it possible to identify different varieties or groups of fish, as well as, the best individual spawners if numerals are used.

Great importance is attached to the techniques of examining the fish during their life time, such as, x-raying, biopsy etc. which have been successfully developed in number of laboratories.

Chapter 18

Methods of Artificial Selection

Objective of Selection

Modern methods of animal selection are inseperably connected with recent achievements in the fields of quantitative and population genetics. Fish selection is based on the planned utilization of genetic variation inherent in all living creatures. An understanding of the nature of genetic variation and the laws by which it is governed is indispensable for the formulation of an effective program of selection work.

In comparison to the breeding of domestic animals and agricultural plants, fish culture is a young science. The domestication of fish and the creation of breeds differing from their wild parents in higher production properties was actually started only a few centuries ago. The only exception is the decorative gold fish (*Carassius auratus*); some gold fish varieties have a thousand year-long history.

The number of domesticated fish species is not great. Besides carps and gold fish, they include crucian carp (*C. auratus gibelio*), tench (*Tinca tinca*), rainbow trout (*Salmo gairdneri*), lake trout (*Salvelinus namaycush*), char (*S. alpinus*), gourami (*Osphronemus goramy*), several centrarchids, tilapia and several other fishes which are bred in some countries of southern and southeastern Asia. Recent years have seen a sharp increase in the domestication of these fishes. This has been brought about by a swift rise in importance of pond fish for the total balancing of fish production. Attempts to reproduce many new species of various freshwater families in ponds and tanks have proved successful. Among these most important are;

1. Silver carp (*Hypophthalmichthys molitrix*), feeding on phytoplankton;
2. Grass carp (*Ctenopharyngodon idella*), feeding on higher aquatic vegetation;
3. Big head (*Aristichthys nobilis*), a zooplankton feeder;

4. Black carp (*Mylopharyngodon piceus*), feeding on benthos, mainly mollusks;
5. Cisco (*Coregous albula*), feeding mainly on zooplankton;
6. Peled (*Coregonus peled*), feeding mainly on zooplankton;
7. Catfish (*Ictalurus punctatus*), feeding on benthos and is also predatory;
8. Pike perch (*Stizostedion lucioperca*), a predator;
9. Buffalo fish (*Ictiobus* spp), feeding on benthos

Domestication of these fishes requires selection to increase their productive properties and adaptation to new conditions of existence greatly differing from natural environments.

American and Soviet fishery biologists claim that selection is applicable, not only to domesticated forms but also to natural populations (Riggs and Sneed, 1959; Nikolsky, 1966). When man is able to fully or partially control reproduction in fish inhabiting natural waters, selection becomes possible. For example, at one of the fish cultural farms in Alaska, USA, all Chinooks (*Oncorhynchus tschawgtscha*) ascending the river for spawning were caught with the help of traps and the best of them were selected for breeding. Young fish were then liberated so that they could grow under natural conditions and subsequently descend into the sea (Donaldson and Menasveta, 1961). The homing instinct driving salmon back for spawning into the river from which they descended precluded the selected stock from mixing with other stocks.

Among the principal object to be accomplished by fish selection are the following (Kirpichnekov, 1966a).

1. To increase the growth rate, by fuller utilization of natural food in ponds and higher consumption of feed (physiological decrease of food expenditure per unit of growth increment);
2. To increase the growth rate, by fuller utilization of natural food in ponds and higher consumption of feed mixture;
3. To increase resistance to oxygen deficiency, to high or low temperature, to higher salinity or other deviations from the normal environmental conditions;
4. To improve resistance to infectious diseases and to infestation with parasites (to develop new breeds resistant to particular disease);
5 To improve the nutritive properties of fish (to increase the caloric content, to decrease the proportional weight of non-edible parts, to decrease the bone content, to increase or decrease the fat content etc).

Sometimes it is important to accomplish other objectives. The aim of carp selection in northern areas is to speed up sexual maturation and to develop the ability to reproduce at relatively low temperatures. In tropical and sub-tropical areas selection is aimed at slowing down maturation, since early maturation is related to the switching over of the metabolism to the development of sex products and may result in uncontrollable reproduction in ponds. There are many more examples of this kind. Particular attention is to be given to selection for combining ability. The utilization of heterosis will become more and more important. The selection of pairs showing maximum heterosis at crossing has acquired a decisive significance in the breeding of carp, tilapia and some American pond fish.

Fish selectionists should also develop breeds adapted to conditions of existence on small reservoirs at high densities, also in cages or aquaria and to feeding on special food mixtures.

The advance of fish culture to countries with a hard cold climate requires an improvement in the cold-resistance properties of pond fish. Adaptations to the ambient temperature may be referred to the group of the conservative specific characters.

According to Ushakov (1963), a change in the temperature adaptation is possible if there is a simultaneous change in the level of cold or heat resistance of the great number of various cell proteins. Therefore the development of breeds having a higher cold resistance (in particular possessing the ability to grow at a relatively low temperatures) is one of the most difficult tasks of fish selection.

Mass Selection: Mathematical Principles of the Theory of Mass Selection

In mass selection, the suitability of each individual is determined by the phenotype, while the genotype characteristics remain unknown. Response in mass selection ® is determined by the general equation (Falconer, 1960);

$$R = I, \text{ delta square minus h square} = S. \text{ h square} \qquad ....(1)$$

where,

S = Selective differential (the difference in certain trait between the individuals selected and the population as a whole);

h square = Heritability of differences (the share of additive genetic variation in the general variation of the character);

I = Intensity of selection

Due to high fecundity of fish, the value of S and I may be much higher than in the selection of domestic animals and poultry. Thus for cattle, intensity of selection is rarely as high as 1, whereas for carps, a value as high as, I = 3 is often obtained. A similar high value of I is sometimes obtained in rainbow trout (Savostjanova, 1969).

In performing mass selection in fish breeding, usually the selection severity coefficient on the rejection rigidity factor (V) is applied, which is calculated by applying the equation

$$V = 100.n/N \text{ per cent} \qquad ....(2)$$

Where n is the number of individuals selected and N is the total number of fish grown. By plotting the intensity of rejection against its severity on a semi-logarithmic scale a curve is obtained. It is obvious that with a decrease in the severity coefficient within the range of 100 to 10 percent there is a sharp increase in the intensity of selection. A further decrease in V (down to 1 percent results in a considerably lower increase in I; and with a decrease in V down to 0.1 to 0.01, the intensity of selection hardly increases at all. Practically for fish with high fecundity, selection gives the best results when the severity of selection ranges from 1 percent (1 : 100) to 0.1 percent (1 : 1000). A further increase of rejection will produce such a low increase in the response to selection that it becomes hardly practicable.

Response to selection is directly proportional to the heritability of the character (h square). In a number of cases this permits a rather accurate estimate of the value of heritability of the character under the selection by applying the equation;

h square + R/S   .... (3)

To obtain such an estimate, selection should be conducted in several successive generations; for in the first generation, the results of selection may be distorted by dominance and epistasis. Heritability determined by the results of selection is called realized heritability. Some investigators believe that this method of estimating heritability is the best when working with pond fish (Moav and Wohlfarth, 1967).

Wlodek (1968) recommends that biometric constants should be used in carp selection for weight. Apart from the mean value, the standard deviation by weight is calculated for each group of fish at all ages. Individuals showing a stable deviation in the course of several years (at the age of 0 +, 1 +, 2 +, 3 + etc.) of a maximum value should be selected for breeding.

Stegman (1968) introduced the concept of " relative growth coefficient" which is calculated by applying the equation;

K = V2- V1 divided by mean X2 – X1   .... (4)

Where V2 and V1 are the final and initial weights of the given specimen and X2 and X1 are the mean final and initial weights of all fish in the pond. Wlodek and Stegman recommended that fish which have the growth coefficient well over 1 in the course of several years should be maintained for breeding.

In applying selection methods, the fact should not be overlooked that in all cases of selection for growth rates, high intensity of selection is of great importance. This can only be realized when a sufficiently large number of fish are grown. Difficulties involved in individual marking of thousands and tens of thousands of fish limit the applications of these methods on a wide scale.

Methods of Increasing Response in Mass Selection

According to equation (1), selection response depends on three values; I, delta and h square. An increase in any of them may bring better results in selection.

Intensity of selection (i) : The value of I may be raised to 3 or 4 in highly fecund fish by increasing the number of individuals grown, and through this, increasing the severity of selection. However, it should be noted that very rigid norms of rejection not only result in an increase in the expenditure incurred in the growing of fish but may involve other undesirable effects;

(a) The weight variation curve is often asymmetric (particularly in the first year of life). Its right part becomes longer due to small number of jumpers (champions), which may not be genetically superior but jump forward only due to a chance (purely paratypic) excess in weight over fish of the same age at the beginning. This weight advantage then increases under conditions of food competition. Heritability of weight differences between the champions and other fish proves to be very poor, and the selection of champions for breeding does not bring good results

(b) Equally undesirable is the so-called correlated response to selection- an often deleterious change in some characters correlated to the character subjected to selection.

One must be very careful in increasing the intensity of selection and pay due regard to a possible decrease in the heritability of the character subjected to selection and to the possibility of the occurrence of indirect deleterious effects due to very rigid selection.

Variability, expressed by the standard deviation, delta, must be sufficiently high. Only genetic variation is important in mass selection. An increase in the value due to an increase in environmental variation is useless for it results in a proportional decrease in heritability. Equally useless is an increase in the share of the phenotypic variance which is conditioned by the non-additive genetic variation.

Heritability (h square) is an important component of the selection response equation. In carrying out selection, all possible methods of increasing heritability should be used.

Heritability is understood to be the relation between the genetic (additive) variance and the general (phenotypic) variance.

Sometimes this heritability is called heritability in the narrow sense of the word, as distinguished from the heritability in the broad sense involving the whole genetic variation irrespective of its nature.

There are two ways of increasing h square;

1. Genetic (additive) variation increases when non-related individuals are crossed. Inbreeding results in a higher homozygosity of the population; whereas out-breeding increases the degree of heterozygosity; *i.e.* increases genetic variation.

   Inbreeding is unavoidable when the number of spawners is low (10-20 individuals) and when the progeny obtained from one or two crossings is maintained for breeding. Such a situation easily occurs in the selection of highly fecund fish, such as carp. In carp breeding, every step must be taken to avoid sharp drop in the abundance of the stock; it must consist at least 50 to 100 spawners. Measures should be taken to have annually a sufficient number of fish for crossing and to make possible the selection of fish from different crossings for breeding.

Particularly dangerous is the so-called heterozygous balance in a population. In this case, the best specimens maintained for breeding appear to be the most heterozygous due to advantages typical of heterozygotes (the so-called over dominance). Genetic variation ceases to be additive, and heritability sometimes drops to a very low value. There are reasons to believe that a considerably high heterozygous balance occurs in carp taken for body weight character, which makes positive selection not responsive. It can be destroyed, and the share of additive genetic variation (delta A square) increased only by means of sufficiently remote crossings.

A new method of increasing genetic variation, which is not yet applied on a large scale, consists in spreading up the process of mutation by new means of irradiation and chemical mutagens. Experiments of this kind was done by R. M. Tzoy in 1967 and preliminary results seem to be promising.

2. Paratypic (non-heriditary) variation may be depressed provided certain conditions are observed, such as,

   (a) growing of spawners under conditions favorable for the maturation of the sex products;

   (b) simultaneous crossings;

   (c) incubation of eggs under conditions as identical as possible temperature, oxygen regime, illumination, water flow etc. must be identical for all the eggs;

   (d) growing of larvae and later of fry, fingerlings and older fish in water bodies rich in food where food competition may not play an important part. It should be remembered that conditions under which fishes are kept should not differ to a great extent from normal conditions; for selection may appear to be effective under certain conditions but give poor results under a different set of conditions;

   (e) simultaneous transference of fishes from one pond to another without mixing fishes grown in different ponds;

   (f) selection mainly at the age which is subject to improvement by selection. Experiments show that in carp the covariance of weight in the first, second and third years of life is not very high. The champions of the first year may not appear to be the best in the second yesr and vice versa. It is not easy to grow a great number of fish of older ages for subsequent selection. A way out is to carry out selection in two or three stages – with low severity in the first year (10-20 per cent) and more rigidly later on, so that only 5-10 per cent of the number of fingerlings selected are maintained for breeding. Thus the summary severity of selection will be sufficient (0.5-1.0 per cent or 1:200 – 1:100).

The observance of the conditions mentioned above will result in a considerable reduction of environmental variance and an increase of heritability. Different fishes may require different methods of reducing paratypic variation.

Correlated Response

A long-lasting, one-way selection for a certain character often results in a change in other morpho-genetically or genetically correlated characters. Indirect changes of this kind often appear to deleterious and result in a reduction of viability and growth rate. This accounts for the fact that many highly productive animal breeds are too delicate. In nature the only criterion of selection is the ability to survive and to produce viable progeny (fitness). In artificial selection, the suitability is determined by the productive properties in which the selectionist is interested, but which, however, are far from being always positively correlated to fitness.

European cultured carp has undergone selection in the course of several centuries mainly for growth rate, exterior characters and scale pattern. Selection has resulted in the deterioration of a number of physiological and biochemical factors, including the blood factor (Steffens, 1964). Here correlated changes have occurred.

An experiment was conducted on the selection of carp for height/length ratio. It appeared that the progeny derived from the highest carp was inferior to other progenies in growth rate (Moav and Wohlfarth, 1967).

A study of the Ropsha carp showed that variations in vertebrae were related to variations in oxygen consumption by fingerlings. There is undoubtedly a correlation here, which means that selection for one of these characters will result in changes in the other.

An example of correlated response which is very important for the theory of selection is provided by changes in size of gonads and in rate of maturation. This relationship has been established for carps and rainbow trout. According to observations, the largest carp often show a slow maturation of sex glands and sometimes a marked decrease in the number of eggs or spermatozoa. Severe selection for growth rate may therefore, have an adverse effect on fecundity. On the other hand, according to Savostjanova (1969), trout which are superior in growth rate to other trout of the same age show an increase in fecundity; at the same time some undesirable effects of high growth rate occur, such as, fatty degeneration of the liver which seems to be a result of increased artificial feeding (Factorovich, 1969). Selection of large specimens under conditions of severe food competition observed at high stocking densities is accompanied by another correlated change, intensification of aggressiveness. In competition for food, the victory is often gained by those fish which are more active in search of food and which snatch it from their neighbours. If differences in growth rate are mainly determined by this factor, then the effect of selection for weight will be very poor.

Examples of correlated responses in selection of non-selected characters in fish are numerous. This phenomenon should not be overlooked in conducting mass selection, for it may reduce all attempts to improve productive properties of a breed by selection.

Genotypic Selection

There are two aspects of genotypic selection, family selection and progeny testing, both of which are applicable in fish culture.

Family Selection and Sib-Selection

In applying family selection, several fish families are grown under identical conditions. By comparing the properties of these families it is possible to determine which of them are the best and to maintain these for breeding. In order to obtain a separate progeny (one family) either one male and one female or a small group of spawners can be used.

Family selection may be very effective. The response equation remains essentially the same as in mass selection;

Rf = if. delta f. hf square   .... (6)

But the magnitude of all terms in the right side of the equation is different. Intensity of selection appears to be lower than the mass selection since it is not possible to grow a great number of families. A reduction is also observed in the standard deviation, which in this case reflects variation in the family means rather than individual variation. On the other hand, the heritability factor becomes much higher, and when the conditions under which all families are kept are equalized it approaches unity.

It is not necessary to cut the fish open in order to be able to examine it, any specimen may be maintained for reproduction. When it is necessary to cut the fish open or when it is impossible to avoid damaging the fish in examining it, the brothers and sisters of the specimens examined from the best families are maintained for breeding. This is called sib-selection.

Family selection methods in fish culture were considered by Kirpichnikov (1966 b). They are as follows;

(i) Crossings – Several pairs or groups of spawners are used to obtain a progeny with the help of artificial fertilization or natural spawning. The number of crossings may be 8-10; and when many aquaria and ponds are available it may be as high as 15 or 20. All crossings are performed simultaneously and the same method of culture is applied.

(ii) Incubation of eggs – When eggs are placed into a Zuger or a Weiss apparatus, or into a crystallizer incubation conditions, temperature, water flow, oxygen content, illumination, density and pattern of distribution of eggs are equalized. The same requirements apply to development of eggs in ponds, but in this case there is an additional requirement of simultaneous spawning in all ponds.

(iii) Larvae – Larvae derived from artificial fertilization are kept in aquaria, cages or ponds at equal densities and under conditions as similar as possible. Larvae derived from natural spawning are transferred to other ponds not later than 3 or 4 days after hatching.

(iv) Communal growing of fish belonging to different families with the aim of testing them.- On reaching the size when they can be marked, larvae can be stocked in communal ponds or tanks for further growing. To decrease competition between separate individuals and to eliminate much of the dependence on stocking weights, the following conditions should be observed;

(a) stocking weights (family averages) are equalized as far as possible;

(b) fish belonging to different families are stocked simultaneously;

(c) dense stocking and starvation should be avoided;

(d) growing experiments are replicated three or four times;

(e) at least 50 fishes from each crossing are used for each experiment (pond).

(v) Separate growing of fishes belonging to different families for testing them. When many identical ponds, tanks or aquaria are available or when it is possible to arrange net enclosures of equal size in a pond, each family is grown separately. The testing of 10 families would require not less than 30 ponds or tanks (when each experiment is replicated in three ponds).

When separate growing is applied, the reliability of results depend on the degree to which it is possible to obviate the differences between ponds or between tanks. The best results were obtained when ponds were partitioned into equal parts with the help of capron netting.

The existence of the so-called material effect (correlation between egg size, survival of larvae and their growth rate) and paternal effect (correlation between the weight of males and the quality of sperm) makes it necessary to grow fishes in testing experiments until they reach a weight of 20 to 30 g, and in large fishes, such as, carp to the weight of 400 to 500 g (Kirpichnikov, 1966a).

The principal disadvantage of family selection lies in technical difficulties involved in the simultaneous growing of many families.

Progeny Testing

There are three ways of progeny testing in fish culture. The first method is testing of pairs without testing males and females separately. The second is to test spawners belonging to one sex (the scheme of testing females) and the third (complete diallele crossing) is the testing of both females and males. All the three methods have been applied in practice.

Pair testing is widely used in carp selection. work in Israel (Wohlfarth, Moav and Lahman, 1961). Progenies of several pairs (up to 15 at a time) are grown together and the combinations are replicated as many as 10 times. The analysis of the results obtained allows the following conclusions to be made;

1. The data on final weight are to be corrected in accordance with the differences I the initial weights (at stocking). For growing fishes weighing 400 to 600 g the correlation factor is 3-4.

2. Comparative experiments on growing fishes of the above weight permit an accurate estimation of the worth of pairs and the selection of the best among them. Results obtained from ponds located in different areas are in good accord.

3. Experiments show that pairs from which the best progenies were derived maintain their superiority when grown separately.

Testing of spawners of either sex was performed repeatedly in carp selection in the USSR. Artificial fertilization was mainly used. The experiments revealed a number of limitations, mainly of a technical nature (the necessity of having more ponds, distortion of results due to competition, difficulties involved in marking and in equalizing initial weights etc.). If all the conditions are observed, which are mentioned in the section dealing with family selection, this method will permit the evaluation of the best and the worst spawners. Two or three males are necessary for the evaluation of females and, correspondingly, two or three females are sufficient for evaluation of males. Experience shows that the number of spawners tested should not exceed 10 or 12, for a greater number of fishes would result in a less accurate evaluation. Improvements in the technique of growing fishes in aquaria may possibly increase the number of individuals tested to 15 -20.

An experiment on diallele crossing of five females and five male grass carp was first conducted by Slutsky in 1967. Twenty five progenies were derived; the investigation was restricted to comparing progenies at embryonic and larval stages of development.

The advantage of progeny testing lies in the fact that it permits the evaluation of separate spawners (or pairs of spawners) and the selection of the best of them for further selection work. Since many fishes are highly fecund, such selection may be of great significance. On the other hand, it should be noted that the great disadvantage of this method is a reduction in the rate of selection. In carp breeding progeny testing requires one or two years (a considerable period of time is also required for other fish). In carp selection this means a slowing down of selection work by 20 to 30 percent. This is absolutely inadmissible in areas (Northern USSR) where carp mature in the 5th or 6th year of life. A way out is either the use of combined selection or growing of spawners in aquaria where water can be heated at any time of the year.

Comparative Response in Mass Selection and in Selection for Relatives

By comparing two response equations

R = S. h square and Rf = Sf. Hf square

It is possible to find out which method is more practicable in a given instance

If S h square is greater than Sf. h f square,

Then mass selection proves to be more effective than selection for relatives (Kirpichnikov, 1968)

The magnitude of the heritability factor (h square) appears to be different for different characters and different populations. Data are available for carp only. The following values of h square have been established;

1. By the weight of fingerlings and two-year olds, 0.1-0.3,
2. By exterior characters, 0.3-0.5,

3. By fat content, 0.2-0.4,
4. By the number of gill rakers and scales 0.3-0.5,
5. By the number of fin rays and vertebrae, 0.4-0.7,

Heritability of weight increases with increasing age and weight of the fish.

In going from mass selection to family selection or progeny testing the heritability factor of weight increases from 0.1-0.3 to 0.8-0.9 and with carefully performed testing to nearly 1. Thus selection for relatives is only advantageous when an increase in h square is not accompanied by a proportionally greater drop in the selection differentials. According to data in family selection of the Ropsha carp, the selection differential decreases by 3 to 4 times. Thus it follows that with h square, mass selection seems to be more useful; whereas with h square less than 0.2, selection for relatives should be preferred.

Chapter 19

Gynogenesis of Fish

Gynogenesis is a rare form of reproduction in which insemination is obligatory, but the sperm nucleus does not take part in development, and the ovum nucleus, with reduced number of chromosomes becomes the zygote nucleus. Therefore only maternal inherited properties develop. It occurs naturally in one form of Crucian carp. This form consists of females only, which propagate by gynogenesis with the help of males of other species, giving only female progeny. The bisexual form is unable to propagate by gynogenesis.

Gynogenesis can not be regarded as usual fecundation which is characterized by amphimixis. It is not identical with parthenogenesis either, though very close to it.

Gynogenesis was first found in some species of free-living nematodes. It is now known to exist also in some other worms, insects, fish and amphibians.

Two cases of natural gynogenesis are known in fish, namely, *Carassius auratus* (Cyprinidae) and in live-bearing, *Mollienesia formosa* (Cyprinodontidae); of which *C. auratus* has been studied more thoroughly because of its importance in the artificial rearing of the species.

Natural Gynogenesis of *Carassius auratus gibelio*

The waters of Japan, China, Korea and the adjacent countries are populated by the species *C. auratus* and in areas further west by its sub-species *C. auratus gibelio* Bloch.

In nature *C. auratus gibelio* occurs in two forms, the usual bisexual form having males and females and the unisexual gynogenetic form consisting females only. There appears to be a certain regularity in the distribution of unisexual and bisexual populations within the limits of natural distribution. Populations in the eastern part of the area are bisexual with a predominance of females, the percentage of males varying considerably. Equal proportion of sexes has been registered only in a few cases. Evidently the numerical predominance of females in bisexual populations results from the mixing of the unisexual and bisexual forms in these populations. The factors determining the percentage of males in these gynogenetic populations are still unknown.

Further to the west, bisexual populations give way to purely female populations. Males are found in these populations as exceptions and do not comprise more than 1 to 3 percent. Such populations are common in Siberia, the Urals and the European part of USSR. In countries of West Europe are found populations of both unisexual and bisexual types, bisexual being rarer.

In mixed populations of *C. auratus gibelio*, gynogenetic females mate with males of the bisexual form of their own species, and, unisexual populations with males of related species. In pond fisheries, offsprings of gynogenetic females are usually obtained with the help of young male carp.

Genetic Analysis of bi-sexual form of *C. auratus gibelio*

Investigations of bi-sexual form of *C. auratus gibelio* was carried out with the population of the Volma fisheries (in the Byelorussian Soviet Socialist Republic), which was brought from Amur in 1948 (Golovinskaya, 1960; Tacherfas, 1966b). It was found that specimens of the bi-sexual form of *C. auratus gibelio*, when mated with each other, yielded equal numbers of female and male offspring.

The diploid number of chromosomes of the bi-sexual form was found to be 94 (by the Japanese scientist Makino in 1939). Maturing of bi-sexual females is accompanied by the usual meiosis. In the last prophase of the first meiotic division all chromosomes are replaced by bivalents. The first division is reducing, and takes place shortly before ovulation ends with discharge of the polar body. As with all fishes studied the ovum of the bi-sexual *C. auratus gibelio* is in the metaphase of the second meiotic division after ovulation contains the haploid number of 47 chromosomes. Shortly after insemination the second (equational) division comes to an end and the second polar body detaches itself. The remaining female telophasic group and the sperm head turn into female and male pronuclei. When preparing for the first division of segmentation both pronuclei draw closer to each other. This process ends with the formation of the diploid metaphasic plate of the first division of segmentation.

Genetic Analysis of Unisexual Forms of *C. auratus gibelio*

Genetic data testifying to natural gynogenesis of the unisexual form of Crucian carp were obtained by the analysis of progeny of females of unisexual populations and males of other species (Table 19.1). In all the crossings progeny consisted entirely of females of *C. auratus gibelio*.

In earlier experiments with *C. carpio* it was observed that 100 Kr and 10 Kr irradiation of sperm produces a most grievous effect on their nuclear apparatus. While 100 Kr irradiation results in complete inactivation of the nuclear apparatus (Romashov *et al.*, 1960). Therefore when the eggs of female *C. carpio* were inseminated

**Table 19.1: Crossings that gave Gynogenetic Progeny of Unisexual Females of
C. auratus gibelio (Silver Crucian carp)**

Sl.No.	Males	Composition of Offspring	Investigators
1.	Wild carp *Cyprinus carpio* L. (S/S n/n)	Silver Crucian carp Female only	Golovinskaya, Romashov, 1947
2.	Linear carp *C. carpio* L. (S/S N/n)	"	"
3.	Leather carp *C. carpio* L. (S/s N/n)	"	"
4.	Carp *C. carpio* L radiation treatment 10 Kr	"	Golovinskaya, Romashov, Tcherafas, 1961
5.	Carp *C. carpio* L radiation treatment 100 Kr	"	"
6.	*C. auratus gibelio* bi-sexual form	"	Golovinskaya, 1960
7.	*C. auratus gibelio* unisexual form	"	Golovinskaya, 1954
8.	Gold fish *C. auratus*	"	Golovinskaya, Romashov, 1947
9.	Gold Crucian carp *Carassius carassius*	"	"
10.	Tench *Tinca tinca*	"	"
11.	Steed *Hemibarbus labeo* Pall	"	Kryzhanovsky, 1947
12.	Roach *Rutilus rutilus* L.	"	Golovinskaya, Romashov, 1947
13.	Loach Cobitidae *Misgurnus fossilis* L.	"	Golovinskaya, Romashov, Tcherafas, 1965
14.	Rainbow trout, *Salmo gairdneri* Rich	Silver Crucian carp Female only	Tcherafas, 1968

with sperm treated with 100 Kr irradiation, the whole progeny consisted of unviable gynogenetic haploids, not counting individual gynogenetic diploids resulting from spontaneous diploidization of the female chromosome complex. With 10 Kr irradiation the effect was lower and the nuclear apparatus of the sperm took part in the development. However, as a result of the fact that the zygote included severely damaged male chromosomes, the whole progeny proved to be unviable, the possibility of individual normal specimens due to diploid gynogenesis as in the case of 100 Kr irradiation, being completely excluded. Similar results were obtained by inseminating the eggs of female *C. auratus gibleio* (bi-sexual form) with irradiated sperm of *C. carpio*. Inseminating the eggs of unisexual females with the irradiated sperm of *C. carpio* gave an absolutely normal progeny as in all other crossings. Consequently irradiated sperm, which is lethal in usual reproduction, produces no grievous effect in natural gynogenesis. Evidently the state of sperm nucleus (from the view point of genetic full-duploidness) is of no consequence in the multiplication of females of the unisexual form.

The above is fully confirmed by the ability of unisexual females to mate with their own males from unisexual populations. As stated earlier such males are very rare. Unlike normal males from bi-sexual populations they are characterized by a lower fecundity and genetic sterility. Evidence of the genetic deficiency of such a male was obtained by crossing it with a female of *C. carpio*. Inspite of the fact that the eggs are of high quality, (which was checked in control crossing of the same female with a normal male *C. carpio*) all hybrid offspring proved to be absolutely unviable, revealing the genetic deficiency of this male. However, for a few years this male gave quite normal gynogenetic progeny when crossed with females of its own kind in natural spawning in pond and in artificial insemination (Golovinskaya, *et al.*, 1965).

Cytological Analysis of Natural Gynogenesis of *C. auratus gibelio*

The exclusion of the male nucleus in the reproduction of the unisexual form of *C. auratus gibelio* which was quite clearly proved by experimental crossings, was also confirmed by direct cytological observations of the sperm nucleus in the plasma of the ova. The first cytological analysis of the process of fecundation of the unisexual form of *C. auratus gibelio* was carried out by K. A. Golovinskaya in 1954. Later these results were confirmed in experiments with females of of the unisexual form from many populations (Lieder, 1955, 1959; Statova, 1963). Observations showed that the centrosome brought in by the sperm doubles, and each of the two centrosomes forms a pole of the spindle of the first division of segmentation. However, during the whole period of preparation for the first division of segmentation the male nuclear apparatus remains as a dense chromatin formation and does not turn into male pronucleus as in usual fecundation. In the first minute after insemination, the head of the sperm penetrates into the plasma and gradually comes closer to the female pronucleus. After that it can be detected in form in the zone of one of the poles of the spindle of the first division of segmentation, and later one of the blastomeres. Lieder managed to trace the head of the sperm up to the stage of eight blastomeres (Lieder, 1959). The further destiny of the male chromatin remains unknown. Evidently it is absorbed by the plasma of the ovum.

By analyzing the maturing process of the gynogenetic females in the populations of the Volma fishery it was possible to ascertain the cytological mechanism securing the permanent number of chromosomes of the unisexual forms of *C. auratus gibelio*.

Chromosomes in the somatic and sex cells of the Volma unisexual females were found to be 141, triploid set of the basic number, 47. Triploidy has also been confirmed by indirect methods, by determining ploidy by the number of nucleoli in the nuclei of the epithelial cells of the fin border of one day larvae and by cytometric data (by the size of cells of some tissues). At present all these indices are widely used for determining ploidy without directly counting the number of chromosomes. Cytometrical investigations have shown that the size of erythrocytes and nuclei of triploid females is on the average 1.4 times greater than of diploid bi-sexual females.

The process of maturation of unisexual females differs considerably from the usual meiotic process, which is evidently closely connected with triploidy. In the late prophase of the first meiotic division all chromosomes are univalent and no conjugation of chromosomes were observed. In the last period of maturation, which ends with ovulation, it is possible to single out several main stages.

1. Stage – I – Concentration of univalents in a limited area of the cytoplasm of the animal pole and formation of an achromatic, multi-pole figure in the zone of chromosomes. This stage is observed immediately after the nuclear capsule is broken and corresponds to prometaphase 1.

2. Stage – II – Formation of a three pole spindle and distribution of chromosomes in three groups at the poles.

3. Stage – III – Transformation of the three pole spindle into a bipolar spindle. At this stage, conventionally corresponding to anaphase 1, chromosomes either distribute themselves into two equal groups or in proportion of 1 : 2 forming haploid or diploid groups.

The above process is abortive and results in no reduction. It ends in uniting all univalents into the triploid metaphasic plate of the only equational division of maturation. It is at this stage that the egg undergoes ovulation. Soon after insemination the division of the polar body takes place. The female triploid complex that remains in the ovum plasma gradually turns into a triploid pronucleus. Further, the process results in forming the triploid plate of the first division of segmentation.

Thus the stable number of chromosomes of the unisexual form of *C. auratus gibelio* of the Volma population is secured by the exclusion of the reduction division. Due to the exclusion of crossing over and reduction which during the usual process of meiosis cause a genetic recombination, the progeny of each gynogenetic female from the Volma fishery is a genotypically uniform clone.

Female from the unisexual population of the Yakot fishery (the Mosco region) have proved to be triploid with ameiotic maturation. It may be presumed that the cytological peculiarities of Volma gynogenetic females are characteristic of other population of the unisexual form of *C. auratus gibelio*.

Experiments to obtain unisexual *C. auratus gibelio* by parthenogenesis gave no positive results. A comparison of the cytology of gynogenesis in *C. auratus gibelio* with data on parthenogenesis of other animals shows that natural gynogenesis of *C. auratus gibelio* is a typical case of triploid apomixes. It may be noted that polyploidy and, first of all, triploidy is a quite common phenomenon under natural parthenogenesis. As in other animals, triploidy of *C. auratus gibelio* is accompanied by the loss of ability for usual sexual reproduction.

The origin of triploidy in gynogenetic females of *C. auratus gibelio* is still not clear. It can be presumed that initially the species had a diploid gynogenetic form, and triploidy came about later as a result of crossing diploid gynogenetic females which produce unreduced gamets with males of the bi-sexual form of their own species or related species. This assumption is based on the results of the well-known model experiments of silkworm, and of some cases of natural triploidy. The assumption is borne out also by the features of the maturation process of unisexual females of *C. auratus gibelio*. The distribution pattern of univalents in haploid and diploid groups during the maturation process of the ovum may be regarded as a result of their cytological and genetic heterogeneity. Therefore, it is reasonable to presume that the genotype of gynogenetic females is of hybrid allopolyploid nature.

While studying the fecundation process of bi-sexual females of *C. auratus gibelio* some instances of spontaneous diploidization of the female chromosome complex, caused by the return of the second polar body into the plasma of the ovum were observed. This process resulted in spontaneous emergence of triploids in the progeny of bi-sexual females. Thus there is a second channel of triploidy of *C. auratus gibelio* which is not connected with hybridization, namely, auto-triploidy.

It is comparatively easy to distinguish females of the unisexual and bi-sexual forms of *C. auratus gebelio*. This is of importance in investigations on mixed populations of the species. Distant crossings and crossing with the use of irradiated sperm were most suitable for distinguishing the two forms. In both cases the difference between the females is quite distinct during the process of embryogenesis, and after hatching, by malformation in the progeny of the females of the bisexual form. Identification of females by the size of erythrocytes is also possible.

Natural Gynogenesis in Fish

Studies of some general features of natural gynogenesis of *M. formosa* and *C. auratus gibelio* reveals the evolutionary importance of this method of reproduction of fish.

Change over from the usual bi-sexual multiplication to parthenogenesis has two main advantages; (i) possibility of somatization during meiosis resulting in preservation of the initial valuable genotype in a number of generations; (ii) rise in effectiveness of reproduction. Both these are true in natural gynogenesis of fish. The assumption that genetic advantages of unisexual lines of *C. auratus gibelio* may be connected with their hybrid constant heterozygous nature and heterosis (which are inherited in a number of generations due to gynogenesis) was made in the first study of natural gynogenesis of this species. Cytological studies has shown that there is no crossing-over and reduction during the process of maturation of ova of gynogenetic females of *C. auratus gibelio* and this supports the above assumption. The above is true for females of *M. formosa*; clone reproduction of this species has been proved with the help of tissue transplantations.

Increasing effectiveness of reproduction under parthogenesis is connected firstly with the doubling of the rate of reproduction due to wholly female structure of populations, and secondly with the removal of the loss of time necessary for meeting of two specimens (a male and a female) under usual reproduction.

For gynogenesis fecundation is necessary, and therefore, the main factor determining the numerical strength of gynogenetic population is the number of males able to ensure the reproduction of gynogenetic forms.

Study of natural gynogenesis of different groups of animals has shown that usually reproduction of gynogenetic females depends greatly on males of bi-sexual forms of the same or related species. This factor limits their population and dispersion, thus depriving them of the main advantages of parthenogenesis. Natural gynogenesis of fish is peculiar in this respect. The results of numerous crossings of unisexual females of *C. auratus gibelio* and *M. formosa* have shown their ability to multiply with the participation of males of various species of fish. In experimental conditions, *C. auratus gibelio* gave progeny when crossed with males of ten different species representing three families. A large variety of species gave gynogenetic progeny of *M. formosa*. It is therefore possible to conclude that in experimental conditions (under artificial fertilization) fish have no limitations in producing gynogenetic progeny by males of different species. Under natural reproduction it is necessary that the ecology of reproduction of gynogenetic females and of the bi-sexual species living with them should be adequate. It is also necessary that the numerical strength of the latter should be great. Evidences of abrupt decrease in the numerical strength of gynogenetic populations of *C. auratus gibelio* were noticed when these conditions were not fulfilled.

Inspite of the limiting effect of these factors, because of their ability to propagate with males of many species, natural gynogenesis of fishes lead to increase in the numerical strength of the species, which is characteristic of natural parthenogenesis of other animals. The marked predominance of the unisexual form of *C. auratus gibelio* within the area of its natural distribution may be attributed to the above considerations.

Spontaneous emergence of gynogenetic offspring in natural conditions was observed under distant hybridization. Hybrid gynogenesis is based on the incompatibility of plasma and chromosomes of the crossing species. Chromosomes brought in by sperm get eliminated to some extent in the primary phase of development of hybrid embryos. Most of these hybrid offsprings are haploids and mosaics with grave malformations. Their development is usually more regular when they resemble mother type. Individual viable specimens emerging among hybrid offspring fully repeat the mother form. Their emergence may be accounted for by complete elimination of male chromosomes and diploidization.

Hybrid gynogenesis proves that in natural conditions mass hybridization facilitates the detection of specimens with hereditary tendency to gynogenesis.

Artificial Gynogenesis

Two unrelated phenomena form the basis of artificial diploid gynogenesis of fish. These are insemination of the ova with genetically inactivated sperm, and diploidization of the female chromosome complex.

Genetic inactivation of male nucleus can be achieved by different means, but ionizing radiation is most often used for this purpose. The quite distinct differential effect of this agent on the nuclear apparatus and cytoplasmatic components of the sex cells allows tha sperm to be treated with very high doses of irradiation. When treated with high doses of irradiation, male chromosomes become genetically inactive, but this does not destroy the sperm's ability to fertilize the centrosome functioning normally to ensure the regularity of the process of the first division of segmentation. Progeny obtained in such crossings consist not only of unviable haploids but individual gynogenetic diploids, as the result of spontaneous diploidization of the female chromosome complex. This phenomenon as been observed in the natural gynogenesis of *C. auratus gibelio*.

The phenomenon of irradiative gynogenesis is known as the Hertwig effect, named after German scientist who first described it. The essence of the Hertwig effect is that when the normal ovum is inseminated with irradiated sperm, the affection of the embryo increases only when the dose is increased to a certain limit. When this limit is achived, the affection decreases noticeably and further increase in the dose of irradiation (up to the limit under which sperms still preserve their ability to fertilize) does not cause further affection of the embryo. Treating with high doses of irradiation produces malformed, unviable offspring with a very few normal individuals. This paradoxical as it may seem, can be explained easily. The effect of irradiation on the embryo increases to a stage where the nucleus of the sperm still retains (if partly) its genetic functions and is able to take part in development. When the dose of irradiation is brought to the limit which causes complete genetic inactivity of the male nucleus, haploid and, in individual cases, diploid rediative gynogenesis takes place.

Characteristics of Gynogenetic Progeny

Most complete data on the characteristics of gynogenetic progeny were obtained from carp reared up to maturation period. All the specimens reared turned out to be females. They were normal in their external appearance but slow in growth. Six out of eight gynogenetic females when crossed with male carp gave normal progeny.

Gynogenetic progeny of Acipenseridae were few, and it was possible to rear the larvae up to the period when they began active feeding. In this period their viability was lower than in the control experiment.

It is possible that the low rate of growth of gynogenetic carp and the low viability of sturgeons are the result of their high homozygosity. Data on the normal fecundity of the gynogenetic progeny have proved to be of great value for selection.

Methods of Increasing the Numerical Strength of Diploid Gynogenetic Progeny

The level of spontaneous diploidization of the female chromosome complex is very low and diploid gynogenetic offspring are very rare. To increase the frequency of diploidization of the female chromosome complex, the method of temperature shock is used. At present this method is used in experiments in ploidy of animals that represent quite different systematic groups, most often of fish and amphibians. Treating the eggs with low and high temperatures in the period of meiotic divisions results in different disturbances in the meiotic process. Diploidization of the female chromosome complex may be caued by the disintegration of the spindle, due to which none of the chromosome sets can form the polar body, or by the return of the polar body into the plasma of the ovum. When fertilized in the usual way, such unreduced ova develop into triploid specimens, and when inseminated with genetically inactivated sperm, diploid gynogenesis takes place.

Investigations have shown that the frequency of cases of induced diploidization of the female chromosome complex greatly depends on the individual characteristics of the female and on the experimental conditions. These most essential conditions are; (i) temperature shock, (ii) the stage at which temperature treatment is applied and (iii) duration of the treatment.

The output of gynogenetic diploids as a rule, varies greatly, being very high under a most favorable combination of parameters.

Temperature treatment gave good results in experiment on *M. fossilis*. The average output of diploid gynogenetic offspring of *M. fossilis* comprises 0.33 percent. When the eggs inseminated with irradiated sperm were cooled at a temperature of 1-3 degree Celsius for ten minutes; after insemination (the stage of anaphase 2) the number of diploid gynogenetic offspring was increased up to 14.9 percent. In the most successful experiments, diploid gynogenetic offspring comprised 50-62 percent of the total fertilized eggs.

In thermal shock experiments when the eggs were warmed up to 34 degree Celsius for 8 minutes after insemination (at the stage of early anaphase 2), the investigators managed to obtain about 17 percent of gynogenetic diploids.

Presently, methods of temperature shock are being sought by experiments on carp. Mastering these methods is of interest for the study of experimental triploidy of fish, particularly for the study of prospects of using polyploidy in fish farming.

References

Andriasheva, M.A,-Commercial hybridization and heterosis in fish culture' Proc. of Seminar/Study Tour in the USSR on genetic selection and Hybridization of cultivated fishes, FAO No. TA 2926, Rome, 1971.

Baskaradoss, K,*et al.* – Cobia (*Rachycentron canadum*) culture, *Fishing Chimes*, 30(8), 2010.

Chaudhuri, H, - Fish hybridization in Asia with special reference to India, Proc. of Seminar/Study Tour in the USSR on genetic selection and hybridization of cultivated fishes, FAO No. TA 2926, Rome, 1971

Chaudhuri, H. – Breeding and selection of cultivated warm water fishes in Asia and Far East, Proc. of Seminar/Study Tour in the USSR on genetic selection and hybridization of cultivated fishes, FAO No. TA 2926, Rome, 1971.

Childers, W.F. – Hybridization of fishes in North America (Family Centrarchidae), Proc. of Seminar/Study Tour in the USSR on genetic selection and hybridization of cultivated fishes, FAO No. TA 2926, Rome, 1971.

Clemens, H.P.- A review of selection and breeding in the culture of warm water food fishes in North America, Proc. of FAO World Symposium on warm water pond fish culture, FAO Fishery Report No. 44, Vol. 4, Rome, 1968

Chalkoo, S.R. – Artificial breeding of trout, *Fishing Chimes*, 30(1), 2010.

Chakrabarty, N.M.- *Ompok bimaculatus* and *O. pabda* comparative morphometric and meristic study of embryonic larval development, Fishing Chimes, 29(6), 2009

Demin, Y.S. – Genetics and its role in biology, Proc. of Seminar/Study Tour in the USSR on genetic selection and hybridization of cultivated fishes, FAO No. TA 2926, Rome, 1971.

Demin, Y.S.,- Mitosis and Meiosis, Proc. of Seminar/Study Tour in the USSR on genetic selection and hybridization of cultivated fishes, FAO No. TA 2926, Rome, 1971.

Demin, Y.S. – Heterosis, Proc. of Seminar/Study Tour in the USSR on genetic selection and hybridization of cultivated fishes, FAO No. TA 2926, Rome, 1971.

Golovinskaya, K.A. – Breeding in fish culture, Proc. of Seminar/Study Tour in the USSR on genetic selection and hybridization of cultivated fishes, FAO No. 2926, Rome, 1971.

Gopakumar, G, G. Syda Rao *et al.*- Silver pompano, A potential species for mariculture in India, Breeding and seed production of Silver Pompano, *Trachinodus blochii, Fishing Chimes*, 31(6), 2011.

Gopakumar, G. *et. al.*– Brood stock development, breeding and seed production of selected marine food fishes and ornamental fishes, Marine Information Series No. 201, 2009.

Hickling, C.F. – Fish Hybridization, Proc. of FAO World Symposium on Warm Water Pond Fish Culture, FAO Fishery Report, No.44, Vol. 4, Rome, 1968.

Honnananda, B.R., N. Basavaraja– Hormonal and environmental manipulation of Maturation and Spawning in Fish, *Fishing Chimes*, 29(4), 2009.

Kirpichnikov, V.S. – Genetics of the common carp (*Cyprinus carpio* L) and other edible Fishes; Proc. of Seminar/Study Tour in the USSR on genetic selection and hybridization of cultivated fishes, FAO No. TA 2926, Rome, 1971.

Kirpichnikov, V.S. – Methods of fish selection, crossing, Modern genetic methods of selection techniques, Proc. of Seminar/Study Tour in the USSR on genetic selection and hybridization of cultivated fishes, FAO No. TA 2926, Rome, 1971.

Kirpichnikov, V.S. – Methods of fish selection- Aims of selection and methods of artificial selection, Proc. of Seminar/Study Tour in the USSR on genetic selection and hybridization of cultivated fishes, FAO No. TA 2926, Rome, 1971.

Konradt, A.G. – Methods of breeding grass carp, *Ctenopharyngdon idella* and the silver carp, *Hypopthalmichthys molitrix*, Proc. of FAO World Symposium on warm water pond fish culture, FAO Fishery Report No. 44,Vol.4, Rome 1968.

Krack, G.V.D. *et al.* – Reproduction; Physiology of Fishes (2nd ed), Edited by David H. Evans, CRC Press, New York.

Kirpitschnikov, V.S. – Efficiency of mass selection and selection for relatives in fish culture, Proc. of FAO World Symposium on Warm water Pond fish culture, FAO Fishery Report No.44, Vol.4, Rome, 1968

Nikoljukin, N.I. – Fundamentals of hybridization in fish culture; Proc. of Seminar/ Study Tour in the USSR on genetic selection and hybridization of cultivated fishes,FAO No. TA 2926, Rome, 1971.

Nikoljukin, N.I. – Hybridization of Acipenseridae and its practical significance, Proc. of Seminar/Study Tour in the USSR on genetic selection and hybridization of cultivated fishes, FAO No. TA 2926, Rome, 1971.

Parazo, M.M *et al.* – Sea Bass hatchery operation, Aquaculture Department, SEAFDEC, Aquaculture Extension Manual No. 18, 1990.

Prem Kumar, *et al.* – Neuro-endocrine control of reproduction and induced breeding through hormonal intervention, *Fishing Chimes*, 32(2), 2012.

Pandey, A.K. and C.V. Mani- *Heteropneustes fossilis* (Bloch), Freshwater catfish, Hypophysial- Ovarian Axis in its Egg Maturation, *Fishing Chimes* 29(4), 2009.

Phadke, G.G. *et al.* – Cytogenetics of fishes, *Fishing Chimes*, 31(2), 2011.

Sneed, K.E. – Some current North American work in hybridization and selection of cultured fishes Proc. of Seminar/Study Tour in the USSR on genetic selection and hybridization of cultivated fishes, FAO No. TA 2926, Rome, 1971.

Syda Rao, G. - Cobia (*Ranchycentron canadum*)-Brood stock development, Success in Induced breeding and also in the larval production for the 1st time at CMFRI, *Fishing Chimes*, 30(1), 2010.

Santhi L. Jemmi – Cobia farming deserves focal attention in India, *Fishing Chimes*, 28(10/11), 2009.

Saxena, Amita and Satesh Vasave- Identification of genetic polymorphism in fish, *Fishing Chimes* 31(2), 2011.

Satyanarayana, Y *et al.* – Genetic markers and their application in fisheries, *Fishing Chimes*, 31(10), 2012.

Sailesh Saurabh *et al.*– Gene transfer technology, Production of transgenic fish, recent development, *Fishing Chimes*, 29(3), 2009.

Tcherfas, N.B. – Natural and artificial gynogenesis of fish, Proc. of Seminar/Study Tour in the USSR on genetic selection and hybridization of cultivated fishes FAO No. TA 2926, Rome, 1971.

Yamaha – Fish community and human society, *Fishery Journal*, No. 41, 1993.

Yamaha – Eels are eaten when they have grown 1000 times, *Fishery Journal*, No.14,1981

Yamaha – Japanese eel aquaculture, *Fishery Journal*, No. 39, 1992.

Yamaha – Pioneering the culture of salt water fishes, *Fishery Journal*, No. 29, 1989.

Yamaha – Sweet fish culture, *Fishery Journal*, No. 38, 1992.

Yamaha – Salmon culture, *Fishery Journal*, no.32, 1990.

Yamaha – Rainbow trout culture, *Fishery Journal*, No. 35, 1991.

Yamaha – Red Sea Bream culture, *Fishery Journal*, No. 33, 1990.

Yamaha – Bastard Halibut culture, *Fishery Journal*, No. 37, 1991.

Index